STEEL

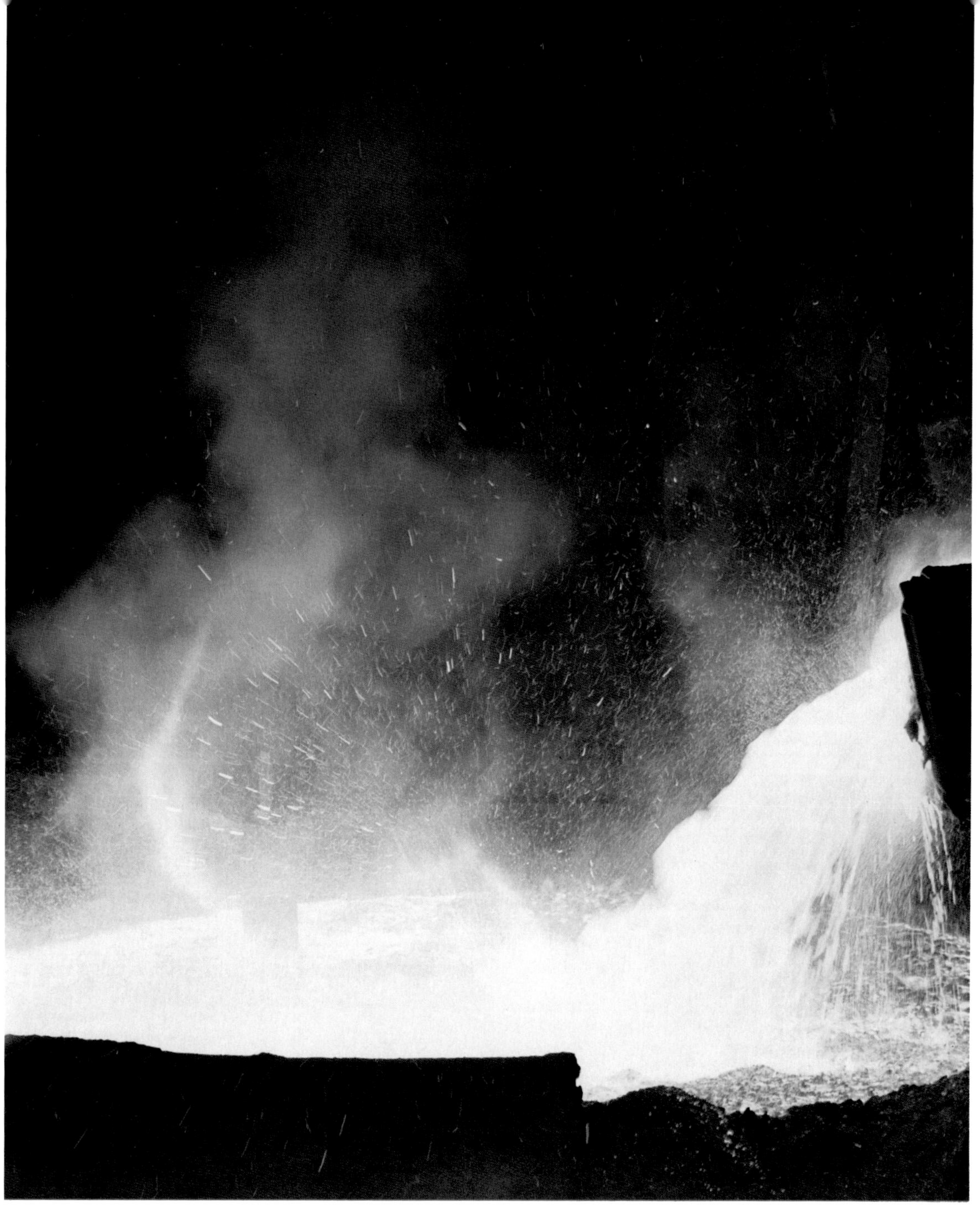

A STUDIO BOOK

STEEL

DAVID PLOWDEN

THE VIKING PRESS
NEW YORK

First published in 1981 by
The Viking Press (A Studio Book)
625 Madison Avenue,
New York, N.Y. 10022
Published simultaneously
in Canada by
Penguin Books Canada Limited

Library of Congress Cataloging in Publication Data
Plowden, David.
Steel text and photographs.
1. Steel—Metallurgy. I. Title.
TN730.P55 669'.142 81-43014
ISBN 0-670-66914-8 AACR2

Printed in the
United States of America

Set in Caledonia

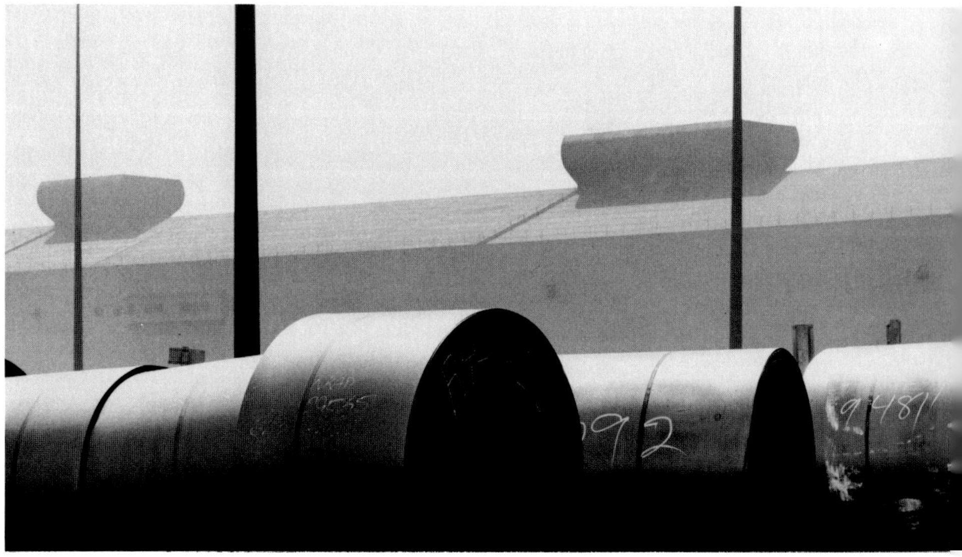

Also by David Plowden

Farewell to Steam

Lincoln and His America

Nantucket
(with Patricia Coffin)

The Hand of Man on America

The Floor of the Sky

Cape May to Montauk
(with Nelson P. Falorp)

Bridges: The Spans of North America

Commonplace

Tugboat

Desert and Plain, the Mountains and the River (with Berton Roueché)

The Iron Road
(with Richard Snow)

Wayne County: The Aesthetic Heritage of a Rural Area
(with Stephen Jacobs)

For Sandra

Acknowledgments

Any book of this nature is not the work of its author alone. Many people have had a hand in bringing this work to fruition, and my sincerest thanks go to them all.

In particular I would like to mention four people whose support was so important. Glenn Hansen spent hundreds of hours assisting me in the darkroom, and I came to rely on his calm expertise so much. Olga Zaferatos, my editor and dear friend, managed to cope with all my disparate manuscript pages, my eccentricities and ideas and concerns, and give me direction. She always encouraged me and gave me faith that the book would work, and in the end she pulled all the pieces together in order to make sure that it did. Her exuberance made the book such fun to do! Michael Shroyer, art director of Studio Books, was so appreciative of my photographs and respectful of my wishes that he managed to indulge me and still choreograph the design beautifully. My wife, Sandra, stood by me through it all and took the full brunt of all my discouragement, my preoccupations in the project that often excluded the family, my exhaustion, and my moods. She had the wisdom and the love to know how to keep me—and the family—together during the two years that it has taken to complete this book. To all four, I cannot express enough thanks.

Contents

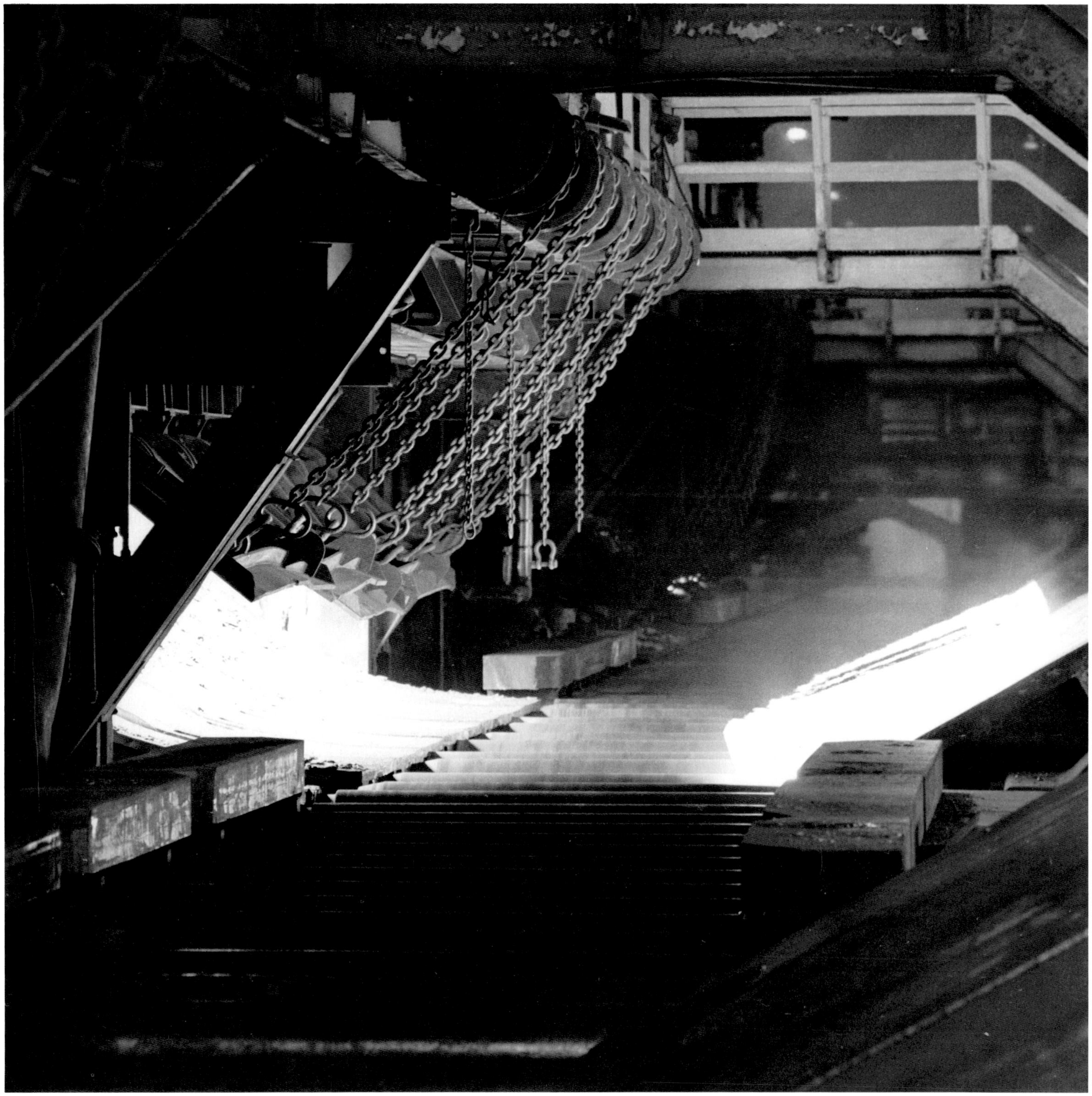

Introduction: An Explanation

This book is about the steel industry as *I* see it. In a larger sense, it is a book about America. The idea for the book began a long time ago—before I knew anything about steel mills or symbolism. It began with my love for machines and for big and utilitarian structures like boats and bridges. When I look back to my youth, it seems that most of my time was spent in engine rooms or in locomotive cabs. When they were not within my exploring range, I would find a steam shovel to scrutinize, or I would prowl around the warehouses and factories of the New Jersey meadows or the Brooklyn, New York, waterfront. I was absolutely fascinated by the industrial world that I was discovering. It had a no-nonsense, brawny honesty about it that I felt did not exist in the more polite surroundings where I slept and went to school. It was clear to me that the other side of the tracks was "where the action was," the place of the forges and machines that had molded our civilization and continuously fed my imagination.

Since I had always headed for the tracks or the waterfront or the mills, it was only natural that I should be drawn eventually to photograph the greatest of all industrial places—the steel mill.

My earliest recollections are of a place filled with great vitality: New York City's East River. My bedroom window looked out onto the river. There, spread before my ingenuous eyes, was the greatest seaport in the New World. Below, on the river, passed every imaginable type of vessel, large and small, from ferries, tugboats, and excursion steamers to freighters, colliers, and tankers with strange flags. The view itself was framed by some of the world's greatest bridge architecture: the Queensboro Bridge to the south was a fantasy of steelwork beyond comparison, and to the north was the strongest steel arch ever built, the Hell Gate Bridge, in front of which looped the cables of the brand-new Triborough Bridge. In the distance there were smokestacks and factories and more bridges, and beyond were the skyscrapers. I remember watching the plumes of steam coming from the tops of the buildings on winter afternoons. I imagined that steam was escaping from safety valves that were there to prevent the city from exploding. Even as a child I saw them as symbols of some sort of latent energy, a power that lay deep within the city's heart, perceived but not yet understood.

Those afternoons were pure Sheeler! Nothing was ever still—the light, the clouds, the river, the city.

I was only seven or eight when we moved away to the middle of Manhattan, and my view became the back side of a church. But soon I discovered a different river, the Hudson, on the other side of town from the East River. Here was another whole spectrum of life to be explored. The world's greatest steamships came to the Hudson's piers, ships that moved majestically, almost regally, and all within view from my vantage point at dockside.

New York was only half my life in those days. Vermont was the other half. Each June my family boarded the White Mountain Express at Grand Central Station for the five-hour trip north, and each September we boarded the same train southward. But to say that Vermont was the other half of my life is a generalization: the other half was the steam locomotive, and at that time Vermont was the place to find steam engines. Most of my cohorts had elaborate electric train sets; in my estimation, their experience was severely limited. Somehow "O" gauge didn't seem the appropriate scale for America, not when I knew firsthand about steam locomotives and ships.

The locomotive, like the ocean liner, was to me the embodiment of contained power, a piece of the cosmos captured in steel. Unlike the liner's, the workings of the locomotive were comprehensible. Constrained within it were primal elements—fire and boiling water and steam. I knew perfectly well that this was a preposterous combination. The whole works might blow sky-high at any moment! Perhaps somewhere deep inside I felt that we were being a bit too presumptuous anyway, trying to harness all these dangerous elements.

In the classroom I was taught that the Golden Age had been attained in Greece thousands of years ago. The implication was that we have been slipping ever since. Nonsense! I knew better. If mine was the age of locomotives and liners and steel bridges, then *this* assuredly was mankind's finest hour.

As soon as I was old enough, I boarded the trains and was pulled by the locomotive back and forth across the country. The railroad became my teacher. My view of America was the view from a train window. I traveled so much that I am sure that I was considered indolent by those individuals driven to more serious endeavors, those more in tune with the puritan ethic—a group that happened to include most of the people that I knew. But I didn't care. I had embarked on a serious endeavor, too—the process of exploring and discovering my country. What I saw became the grist out of which grew so many photographic projects. My head is still full of visions that were plucked from along those hundreds of thousands of railroad miles. One in particular stands out most distinctly: my first glimpse of a steel mill.

It was in 1945 when, as my train sped by, I caught only a fleeting glance of a steel mill, but to me it was a scene unlike anything else that I had encountered before.

Staccato glimpses of steel mills became exclamation points in all my subsequent journeying. They possessed the same kind of symbolic power that I associated with the locomotive and the liner; they were evidence, reassuring signs of a great strength that lay within the land. Seen at a distance from a speeding train, the mills were incomprehensible. It was impossible to see what was happening there or how anything worked. This is not surprising, as I later learned, because steel mills are such vast and complex affairs that even the steelworkers themselves are largely unfamiliar with any but their own particular segment of the work.

But before I had a chance to discover more about my land, I was derailed. My own latent puritanism finally rose to the surface. It was time to be serious about life, to stop exploring and start preparing myself for life. I went to college and I was bored, but my love of railroads and things exciting refused to die. This love had to be justified now. I decided that the best course would be to become president of the Great Northern Railway. I became an economics major so that I would be prepared to take up the reins of executive power if I should be called upon to do so. I was, to my great surprise, tapped by the Great Northern. I was not sent directly to the president's chair, but, much to my happiness, to an outpost in the cornfields of Minnesota, where one of the last stands of the steam locomotive was taking place.

Then the engines were retired and simultaneously I was promoted up the first rung in the ladder toward the executive suite. I saw myself faced with a life of sitting behind the desk, not riding the rails, so I quit and went back to the life of indolence. This time I went to Europe. When I returned home on the *Queen Mary,* one of the last of the great liners, everything seemed changed, or at least I was aware that change had taken place.

It was a little after mid-century when I stepped down the gangplank into a seething, frenetic, energetic, confusing world—America. Then, for the first time, I began to understand Frederick Jackson Turner, whom I had read in history class at Yale. Turner came as close as anyone to perceiving the nature of the American spirit. He believed that the existence of the frontier was the key factor in the development of American society. By his definition, the frontier was not solely a geographical line; it was also a process that took place there, by which society was continually being broken down and built anew. In the course of these changes emerged those character traits that have become synonymous with America: self-reliance, ingenuity, determination, and self-confidence. The pioneering spirit that we Americans have always revered and with which we have become identified is at once our great conceit and our strength.

When Turner spoke in 1893, the American frontier was, in its classic sense, all but closed. There were other frontiers on the horizon, however; there were frontiers

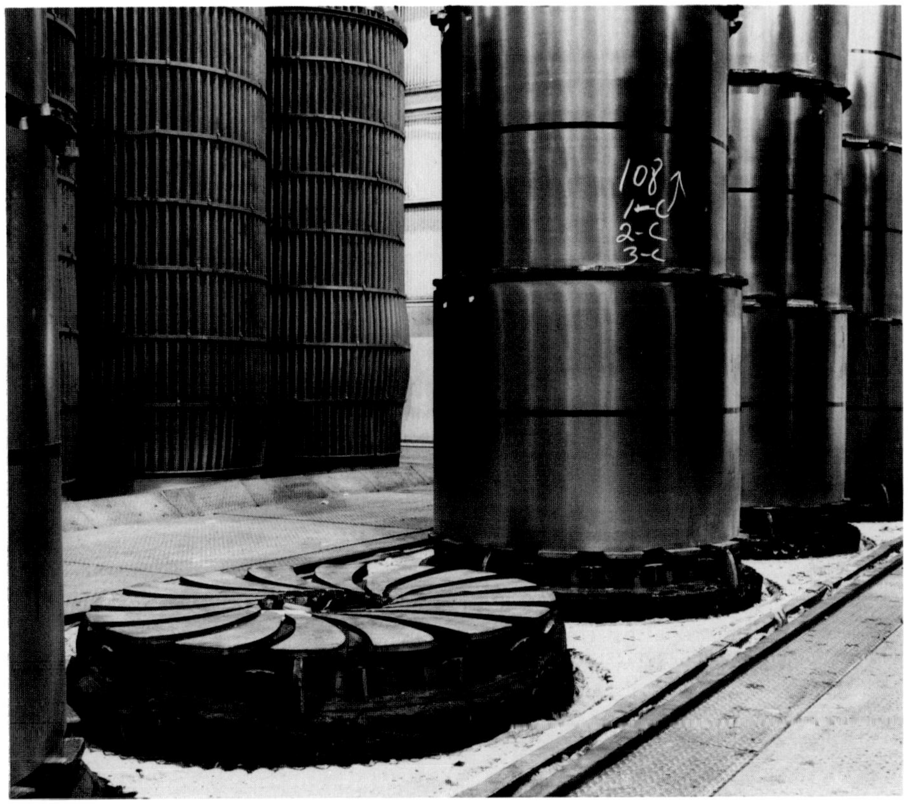

where the opportunities to exercise Manifest Destiny seemed every bit as limitless—or so we believed. The Industrial Revolution was in full swing. "Things," as Emerson wrote, "were in the saddle." This promised to be an age of triumphs on the mechanical frontier. The pioneering spirit was there; America's hands were still callused. We had simply put down the ax and grasped the lever. We directed the same energy and resourcefulness with which we had conquered one frontier to work on others.

The machines and industrial architecture of the earlier

generation of mechanization reflect this pioneering spirit. They are full of exuberance and often quite unsophisticated by today's standards of specialization. But their boldness and directness make them, I feel, that much more interesting. Their importance is obvious because they *look* important. Much of today's technology is more cerebral than physically apparent. It does not as often produce palpable objects—a Brooklyn Bridge, a locomotive, a Chrysler Building—three-dimensional forms that speak to us directly and eloquently of human endeavor and genius. They are works of strong motivation and feeling. Conversely, the new technology seems singularly dispassionate. It has evolved in a more intellectual sphere; it has reached such degrees of complexity that only the initiated—the specialist—can understand its implication, let alone appreciate it.

So often the tangible evidence of this new technology—the magnificent rocketry of the space age, for instance—is sealed off by physical barriers, so that it is inaccessible to, and unseen by, most people. In that sense, today's technology is exclusive, as remote from us as the contrails across the sky.

I think most people want to rejoice in the achievements of their country, whether this is called pride or patriotism or chauvinism. Rather than fostering our pride, technological specialization has confounded most of us; it has not really enabled us to feel that we have taken part in the collective accomplishments of this nation. It seems that the more intangible the products of this new technology, the greater becomes our need to seek symbols. For some people the symbol may be the flag, for others it may be the Statue of Liberty or an august personage who embodies the spirit of a nation and its aspirations. For my part I have looked to the iconography of industry—to locomotives, liners, bridges, and steel mills. These, to me, are the most expressive American symbols.

This is the reason why I chose to photograph the steel mills. I came to know the mill better as I worked; the place was just as awesome as I had remembered. But my frustration grew as I tried to digest the essence of the place. How presumptuous I had been to believe that it was possible to record with a still camera the vitality of a steel mill! A steel mill is the very antithesis of stillness. Most of what I saw and sensed eluded my grasp: all that overpowering energy, the barrage of stimuli, the cacophony, the heat, the smells, the overwhelming machinery—the mind-boggling complexity of it all was not within the ken of the camera.

The book is organized to follow the sequence of processes that transforms iron ore into finished pieces of useful steel. It does not attempt, however, to document each and every step in this metamorphosis; indeed, there are certain processes that have been omitted entirely. Many of steelmaking's functions are, from a visual point of view, so similar that to include each one in a book of this nature would have been redundant and would have broken the visual momentum of the book, a momentum that attempts to parallel the drama of steel-making. For this reason I had to exercise a sort of photographic triage. There are photographs of the rolling process, but not of cold-rolling, or blooming, or slabbing. There are no photographs of strand-casting, nor any of an electric furnace, nor any of forging, pickling, and extruding. I am fully aware that these are all extremely im-

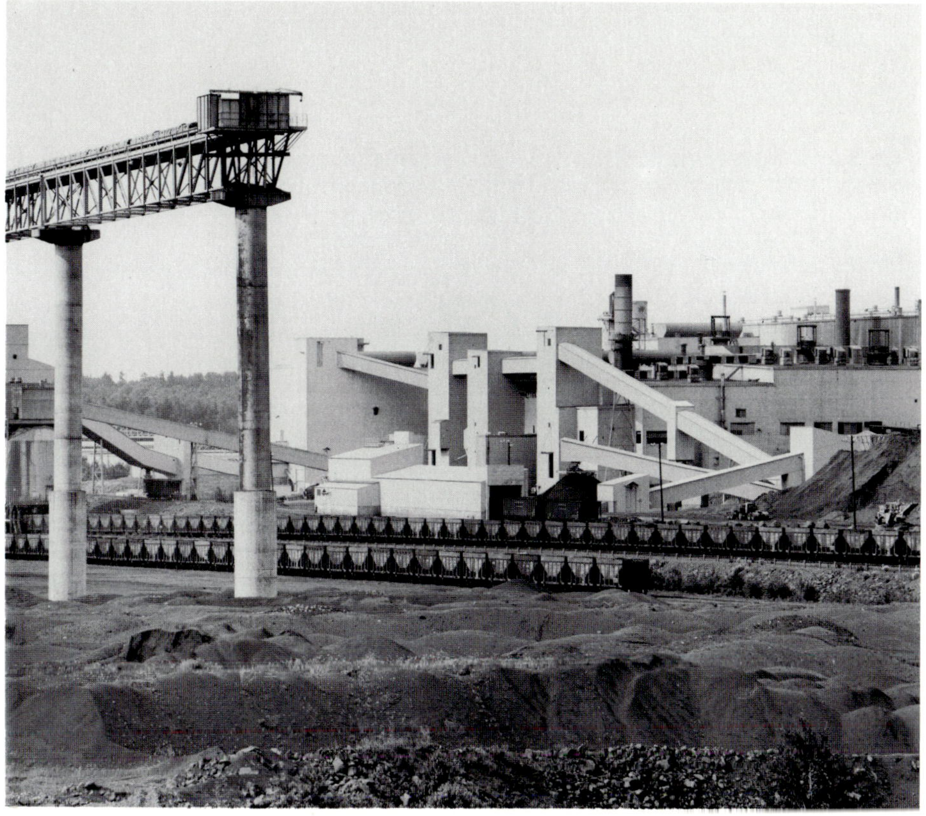

portant functions. But because parallel or similar situations made better photographs, I left those steps out in order to strengthen the overall character of the book.

I discovered—I am sure to the dismay of the steel industry—that some of the newest and most innovative procedures in operation would have made uninteresting photographs, simply because the camera could not have done justice to their importance. For instance, a computer console represents an astounding technological achievement, but visually it is not exciting.

Steel-making is a process that requires many hands, and in this area I have also excluded a great deal so as not to be redundant. Photographs showing "man-at-work" all have a tendency to look alike; photographs showing a worker standing next to a machine, or sitting in front of a control panel, or discussing business with another manager give no clue whatsoever as to the actual relationship between the people, the tools they work with, and the significance of their jobs. For these reasons I purposely photographed only a few situations in which workers were included, choosing only those where the relationship between man and machine was obvious and unique to the steel industry.

All in all, the selection of images was made primarily by visual criteria, and the book is to be taken as an impression, as my view of the steel industry.

Blast-furnace computer system.

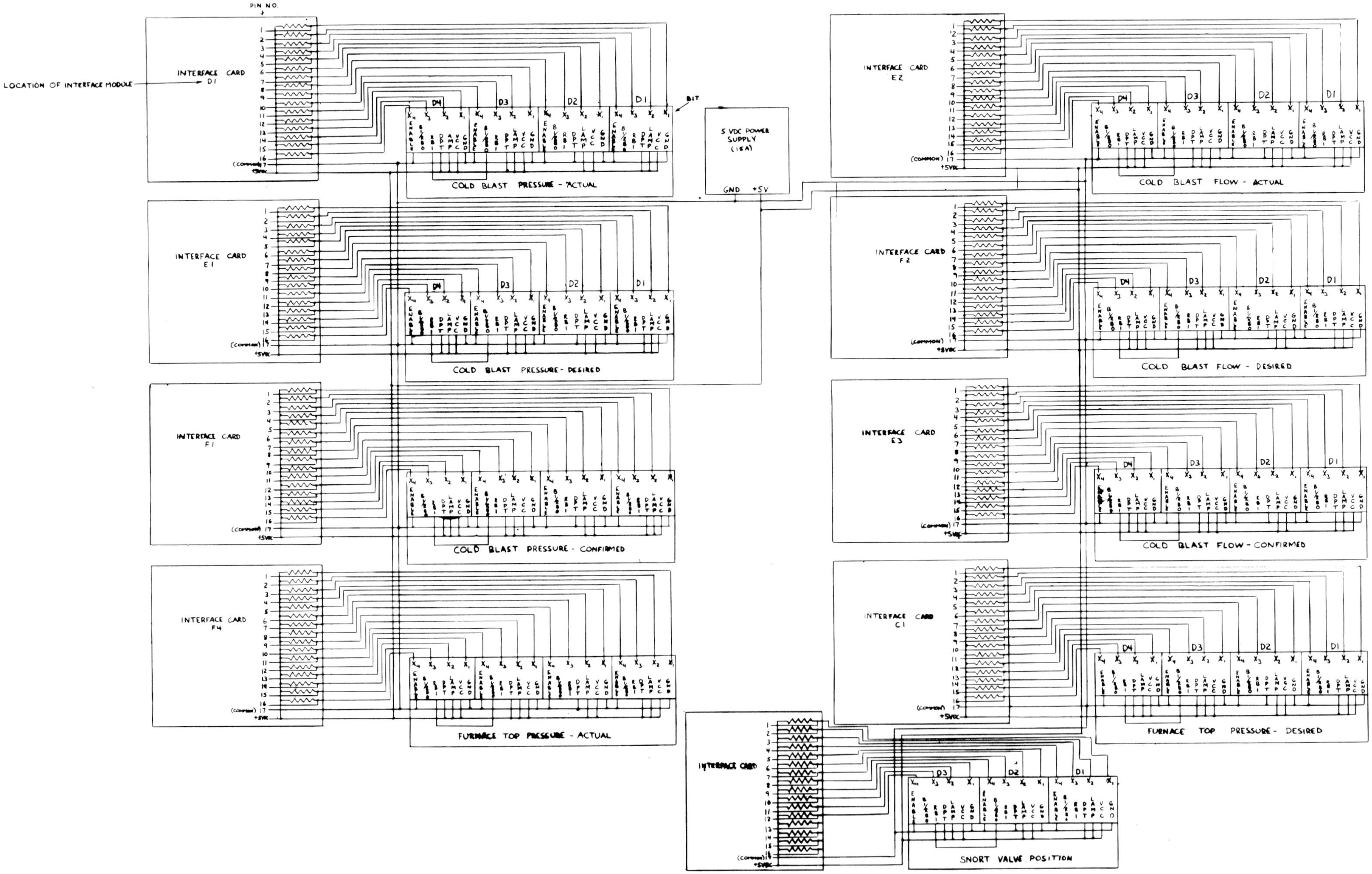

Steel, "The Giant Among Metals"

Steel does not exist in nature. It is an alloy, essentially refined pig iron—known also as "blast furnace metal" or "hot metal" to the steelmaker. Steel-making, therefore, is a refining process whereby certain elements present in pig iron, such as carbon, are reduced. By strict definition, steel contains less than 1 percent carbon, which makes it less brittle and more malleable than its parent metal. The great advantage of steel is that pound for pound it is stronger than iron. This means that it is possible to use far less material in any given situation than if iron were employed. A piece of iron comparable in strength to one of steel would weigh more. It would also have to be larger, and much of its strength would be used to hold up its own weight.

Despite steel's advantages, it was not initially embraced as a structural material. Steel was first successfully used in America as a building material in the construction of the Eads Bridge, which was completed across the Mississippi River at St. Louis in 1874. However, it was not until the last decade of the century that steel supplanted iron as the primary structural material.

For millennia, iron- and steel-making had been comparatively rudimentary affairs. Usually, a local enterprise situated near the source of material produced small quantities of iron and steel to serve the needs of an unmechanized, largely agrarian society. Then, in the middle of the last century, a series of events irrevocably changed the nature of steel-making and the world as well.

One of the most important of these events occurred in 1845 with the discovery of what proved to be one of the richest and largest bodies of iron ore known to man. The iron ranges of the Lake Superior region of North America provided the raw material that was so instrumental in contributing to America's preeminence. The opening of the Soo locks on the St. Marys River between Lake Superior and Lake Huron enabled this wealth of ore to be shipped directly to the lower lake ports, where it could be fed into the furnaces of America's burgeoning steel industry.

More important to the industry as a whole was a combination of events that made it possible to produce large quantities of steel economically. The first of these occurred in 1847 when the Englishman Henry Bessemer perfected a pneumatic method of producing steel. Bessemer employed a vessel—a converter—in which air was blown through molten iron. This removed the impurities that caused iron to be so brittle and thus made steel. The identical process was developed independently in America at the same time by William Kelley. Bessemer received the patent first, and the process bears his name. The first practical application of the process in America occurred in Troy, New York, on February 16, 1865, at the works of Winslow, Griswold, and Holley.

A single process in itself does not guarantee the success of a whole industry. It usually takes someone to put all the pieces together, and in the case of the American steel industry that person was Andrew Carnegie, one of the era's great entrepreneurs. Carnegie, with the help of the engineering genius of Alexander Lyman Holley, of the Troy firm, took the Bessemer process and made it into an entire system for making steel.

In 1868 an even better method of steel-making was introduced—the Siemens-Martin process, known to the world as the open-hearth furnace. This innovation made the production of even greater amounts of steel possible, and it soon supplanted the Bessemer process as the principal method of making steel.

There were other factors contributing to the success of steel-making, of course. The coke industry had advanced to the point that it could produce a fuel with which the iron ore could be smelted. Coke would replace the charcoal that was used in the smelting process, and the smelting process itself was perfected. Rolling replaced forging as the principal method of working steel, which made it possible to produce vast quantities of finished steel economically and in a great variety of shapes. A fleet of boats was built to carry the ore down the lakes, and a network of railroads was developed in order to take raw materials to the mills and to carry the finished product to the marketplace.

The American steel industry has evolved from that time into a vast and complex organism which, like the country itself, defies simple description. Ironically, steel's tremendous importance is too often overshadowed by the extraordinary exaggeration that is used in describing it. Descriptions of steel are full of superlatives; statistics about steel deal with unbelievable figures. Steel is "as vital to modern civilization as air and water are to life." It is "the giant among metals"; "the basic metal of civilization"; "the most important metal in a metallic age"; "the master metal." The truth is that steel *is* all that the descriptions promise, and much more.

One of the most florid statements of steel's importance comes from a turn-of-the-century edition of one of the most practical of works, *The Making, Shaping and Treating of Steel*:

> *Thus, from our towering skyscrapers, our massive bridges, and our immense ships, where as great beams, cables, and plates, it supports loads almost greater than the mind can conceive, we can trace it even to our parlors, where as invisible hairpins, it may support milady's tresses and, as the strings of her piano, it sends forth at her magic touch sweet sounds of melody.*

Steel-making is an industry demanding of hyperbole, for it is one of unparalleled size, where processes are measured in terms of millions of tons and thousand-degree temperatures. Ninety-five percent of the world's metal-tonnage production is either iron or steel. Forty percent of America's industrial work force is engaged in either the manufacture of steel or its use in finished form. It really is not one industry, but a combination of many, each with its particular methodology, special technology, and machinery. All these enterprises are integrated into a single system with a singular goal: the manufacture of steel.

Raw Materials: Iron Ore

Steel-making is a Bunyanesque occupation. It is a business that deals constantly with the problem of moving huge amounts of material from one processing center to another. It is said, in fact, that 90 percent of steel-making involves the handling of raw material, and that no other manufacturing process in the world uses as much as the steel industry.

The drama of steel-making all begins in the earth's crust, in the solidified residuum of its resources, with rock—hematite, taconite, magnetite, dolomite, limestone—and with coal.

Iron, the mother of steel, happens to be one of the world's most abundant elements; it makes up about 5 percent of the earth's surface. Yet, in spite of its abundance, iron is not always present in a useful state. Unlike some metals that are found in their pure form, and with the notable exception of meteorites, iron is always found in combination with other elements. These "other elements" are the bane of the iron and steel industries, for the iron must be separated from them in order to be useful. The medium in which iron is found is pragmatically referred to as "iron-bearing material." But it is another matter altogether whether this material can also be defined as "iron ore." The differentiation depends entirely on whether the iron can be extracted from the unwanted elements at a profit. Ore, then, is an economic term of flexible definition.

Nature was very generous to America when the world's deposits of iron were allotted. Some of the richest and largest deposits were discovered in 1845 around the shores of Lake Superior. Over four-fifths of our domestic iron has come from these ancient ledges in Minnesota and Michigan, from iron ranges with such exotic names as Mesabi, Marquette, Menominee, and Vermilion.

Without question, these deposits were "ore." The material was rich, some of it consisting of more than 50 percent iron. Moreover, most of it was deposited close to the surface; about 90 percent of the ore was readily accessible and could be mined by the open-pit method. Steam shovels were used to strip away the overburden—the forest and the topsoil—to expose the ore. Once exposed, it was scooped up and hauled by railroad to the lake ports where ore boats were waiting to carry it away. Billions of tons of iron ore were extracted from the earth in this manner to feed the insatiable demands of the burgeoning steel mills and a growing nation of consumers.

Until recently we Americans generally have not been overly preoccupied with the idea that our resources might be finite. The coffers of the iron ranges were just as "inexhaustible" as the white-pine forest had been—the forest that had covered the same landscape for generations before the miners arrived. By the middle of the twentieth century, however, two world wars and several hundred million automobiles, refrigerators, and bridge girders later, the rich ore was gone. We had to look to other countries—Canada, Liberia, Venezuela—to meet our demands. But we also searched for new sources and soon discovered that there was still a lot of iron "by the shining big sea water." It was locked up in a hitherto useless rock called taconite.

Taconite is not the ideal rock to begin with. It is extremely hard and dense, and the ratio between waste and iron is wrong: too much waste, too little iron. The trick was to turn taconite into an ore. This called for the creation of a new industry, one which would "beneficiate" the quality of the taconite so that it could be fed into the blast furnace in place of the high-grade ore.

Like most high-grade ore, all domestic taconite is mined by the open-pit method. Because of its extreme hardness, however, taconite must be blasted from the mine walls and crushed before it can be processed. In fact, it is so hard that jets of flame as high in temperature as 4,300 degrees F. must often be used to "drill" the blasting holes. (Today high-speed rotary drills are replacing the jets of flame.) Once shattered, the shards of rock are crushed and ground in special mills until they have been pulverized into a fine powder from which iron particles can be extracted. The extraction is accomplished by either magnetic separation or flotation, depending on the type of ore. In flotation, reaction to chemical reagents causes the iron-bearing particles to sink, while the waste material remains suspended and is skimmed off the top. After the water used in the process is filtered out, the iron concentrate, as it is called, is delivered to the balling mill. There it is mixed with a binder and formed into soft "green" pellets about the size of marbles. These pellets are then fired in a kiln or on straight grates until they have hardened enough to be easily transportable.

The pellets that are produced by the beneficiation process contain from 60 to 66 percent iron, more than most natural ores. Because they are round and nearly uniform in chemical composition, the pellets make better and more stable charges for the blast furnace; consequently, the iron that is produced is of a more consistent quality. The process of beneficiation requires an enormous investment, however, and for every ton of iron recovered from taconite, two tons of waste are produced. In some locations the disposal of this waste has been of great concern to environmentalists and mining companies alike.

Virtually all pelletizing in Minnesota, Wisconsin, and Michigan takes place in plants near the mine site, and the process today has become an integral part of iron-making itself. Pellets are transported in the same way as the high-grade ores were—by railroad to the ore docks of such port cities as Duluth, Minnesota; Superior, Wisconsin; and Escanaba and Marquette, Michigan. There they are loaded into the holds of those vessels indigenous to the Great Lakes, the bulk-cargo carriers known as "lakers." The laker, like everything associated with steel-making, is huge. A new species, the self-unloading laker, is a thousand feet long and capable of carrying sixty thousand tons of ore. Unfortunately, its efficiency is not matched by its appearance. This behemoth and her small sisters nonetheless manage to keep America's blast furnaces well fed with pellets. Some sixty million tons of domestic and imported pellets are consumed by the industry each year.

Open-pit iron-ore mine.

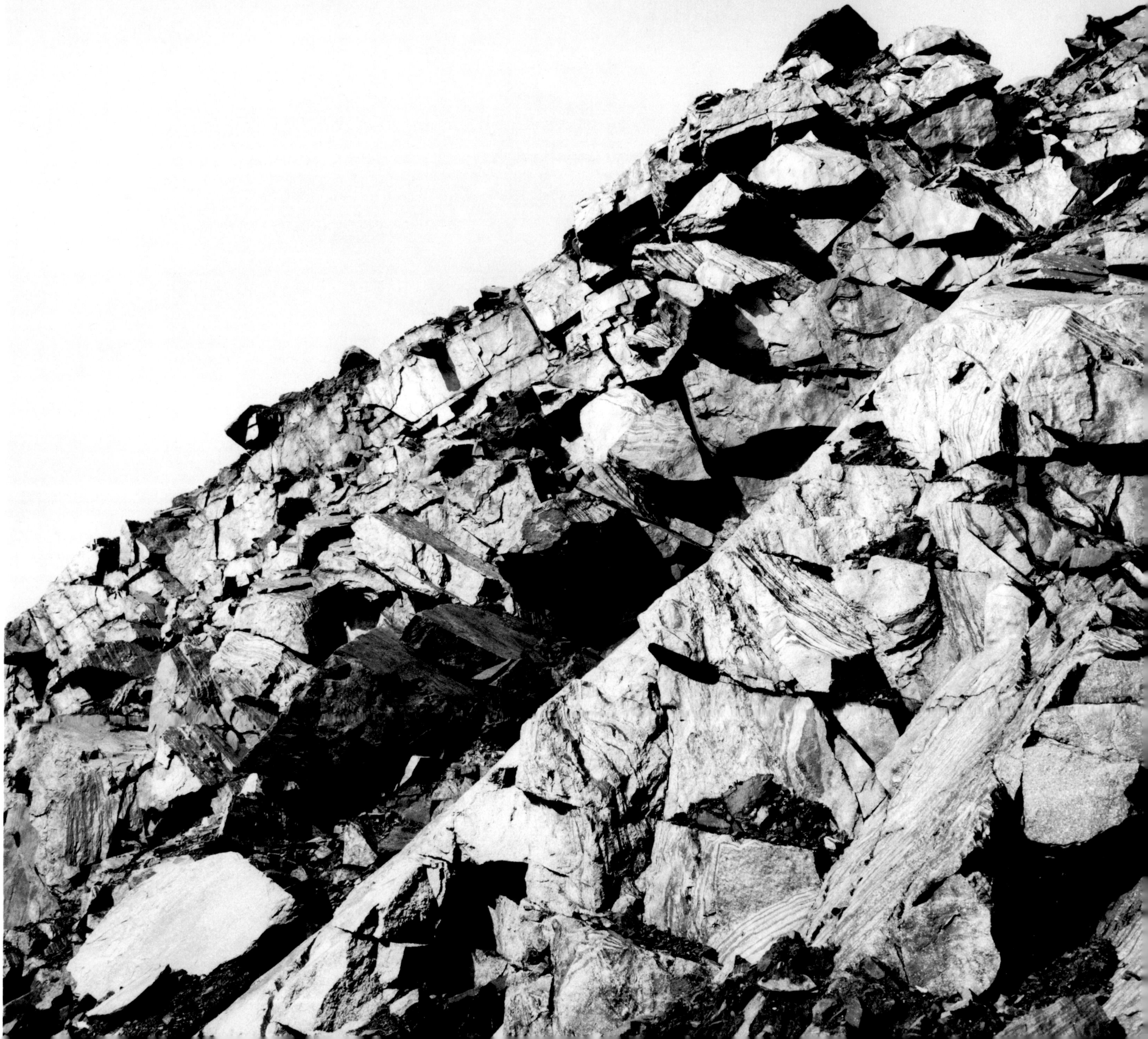

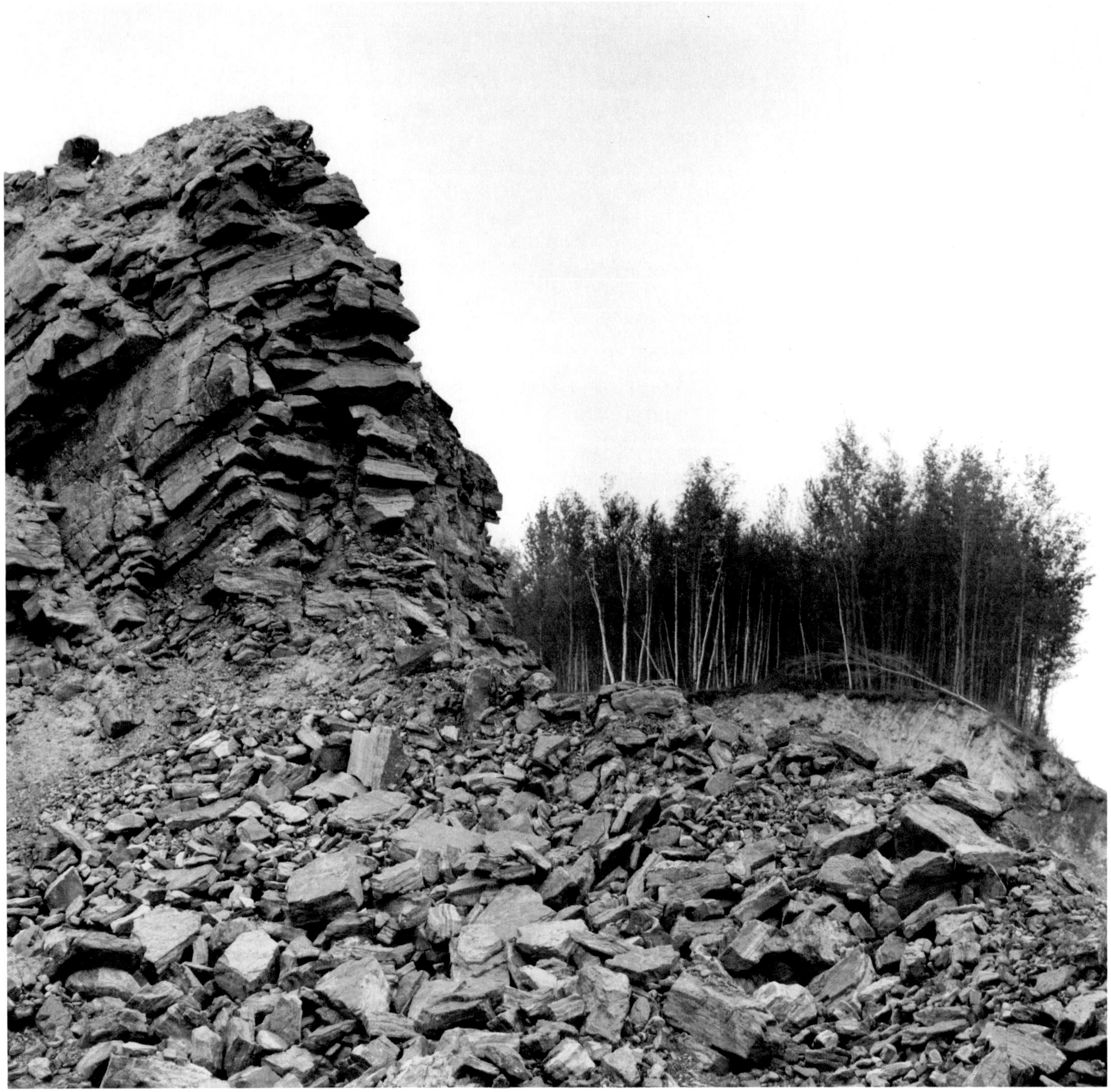

Open-pit iron-ore mine.

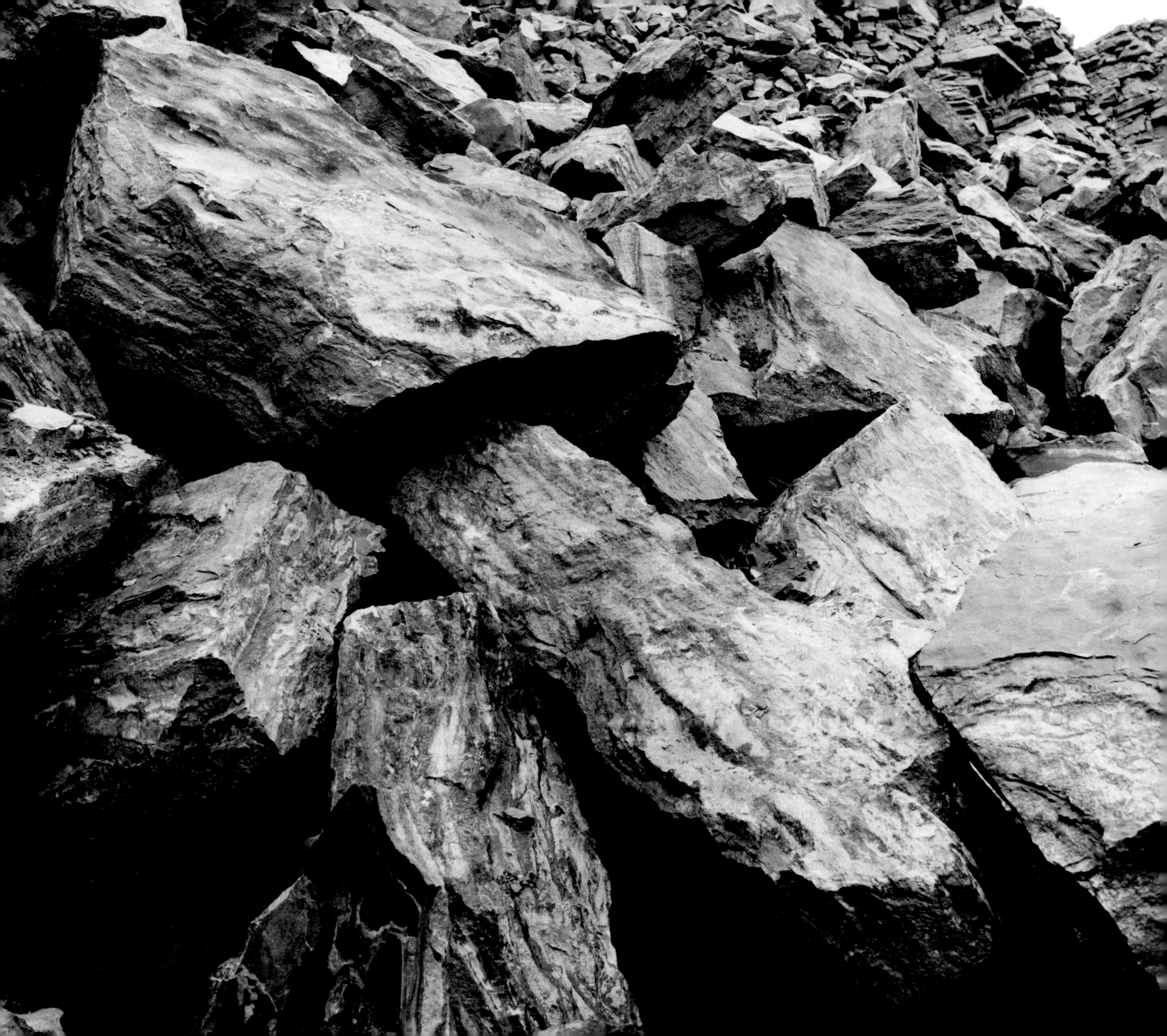

Open-pit iron-ore mine.

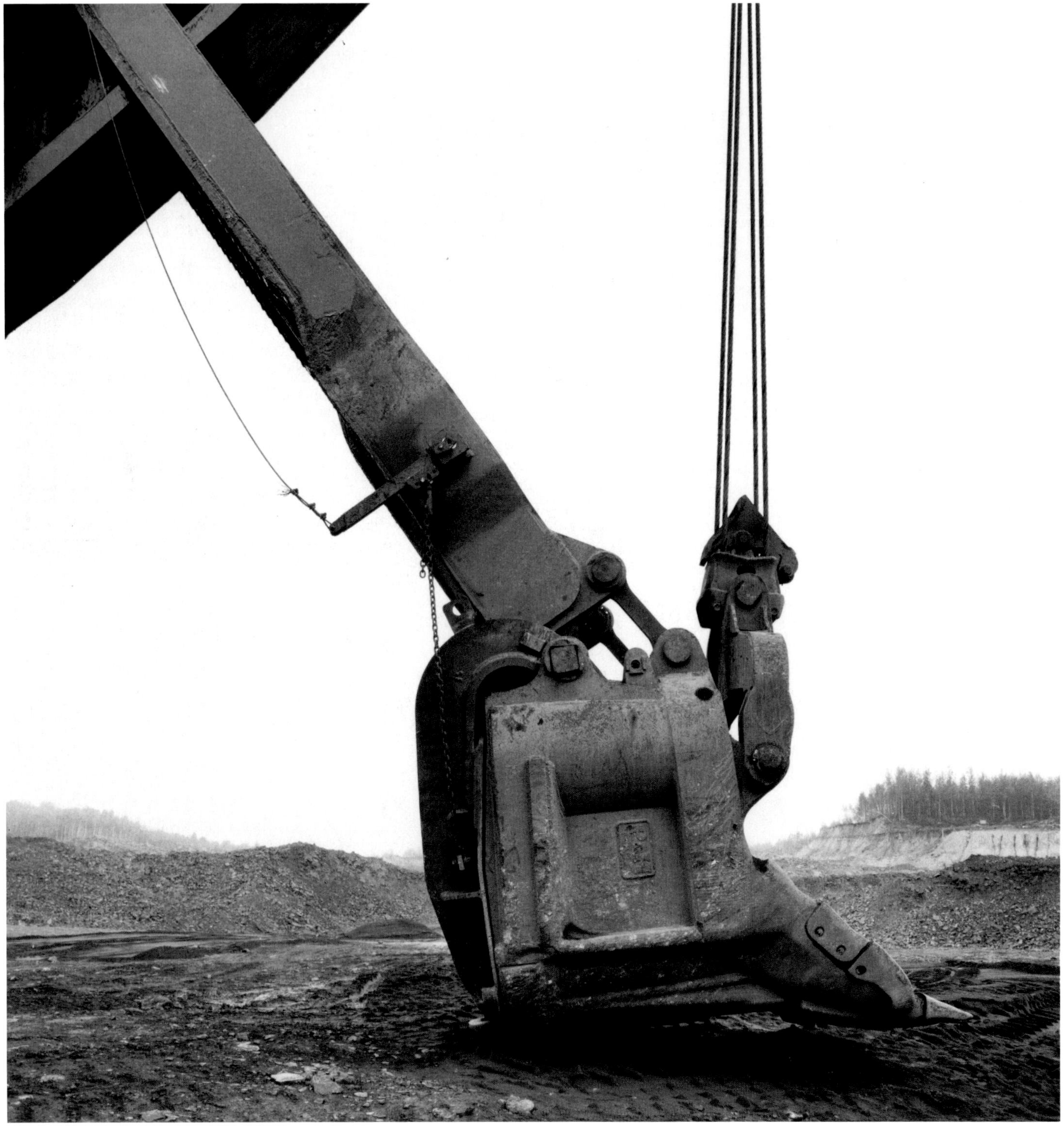

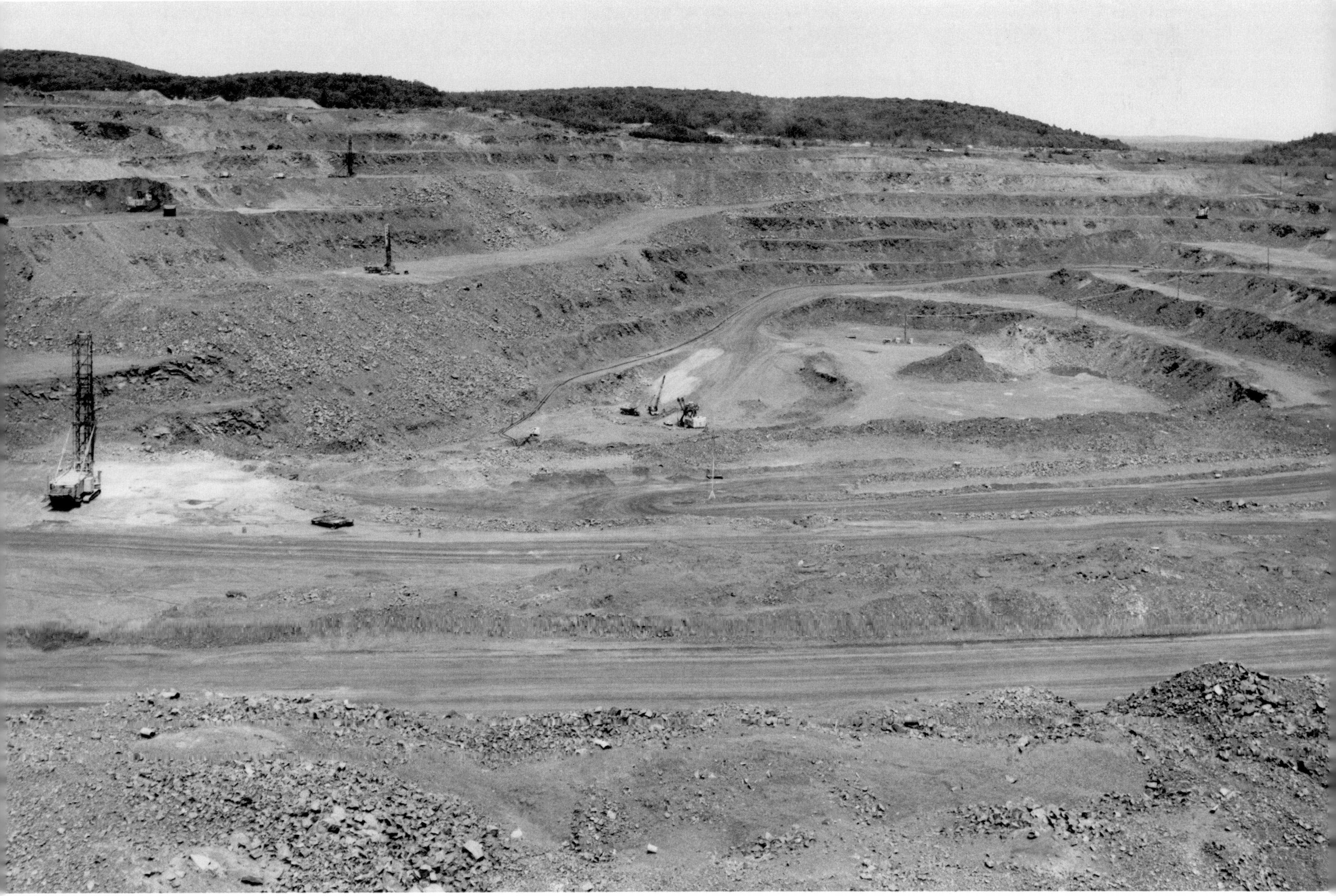

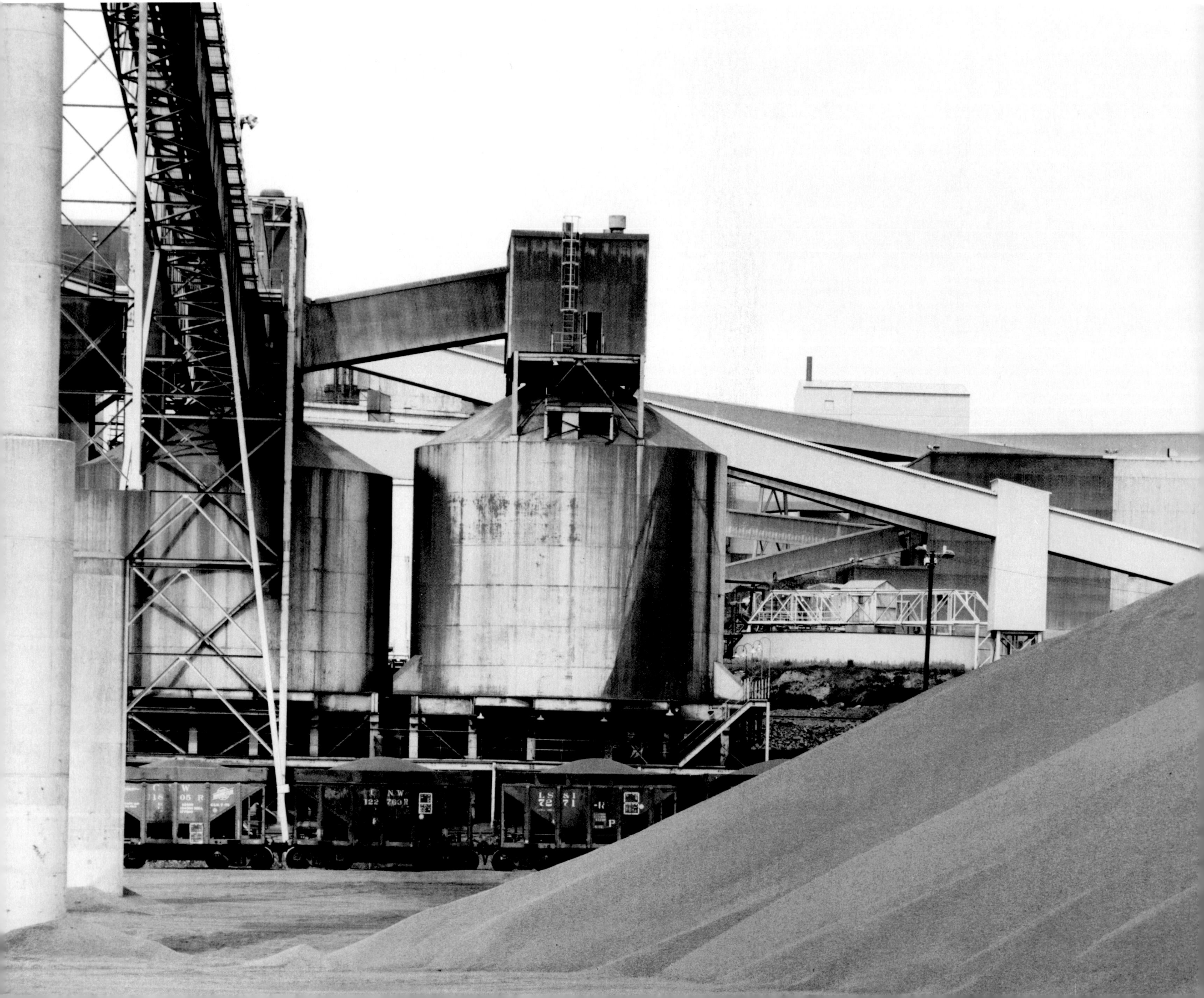
N W
122 763 R
LS &I
P

Taconite plant.

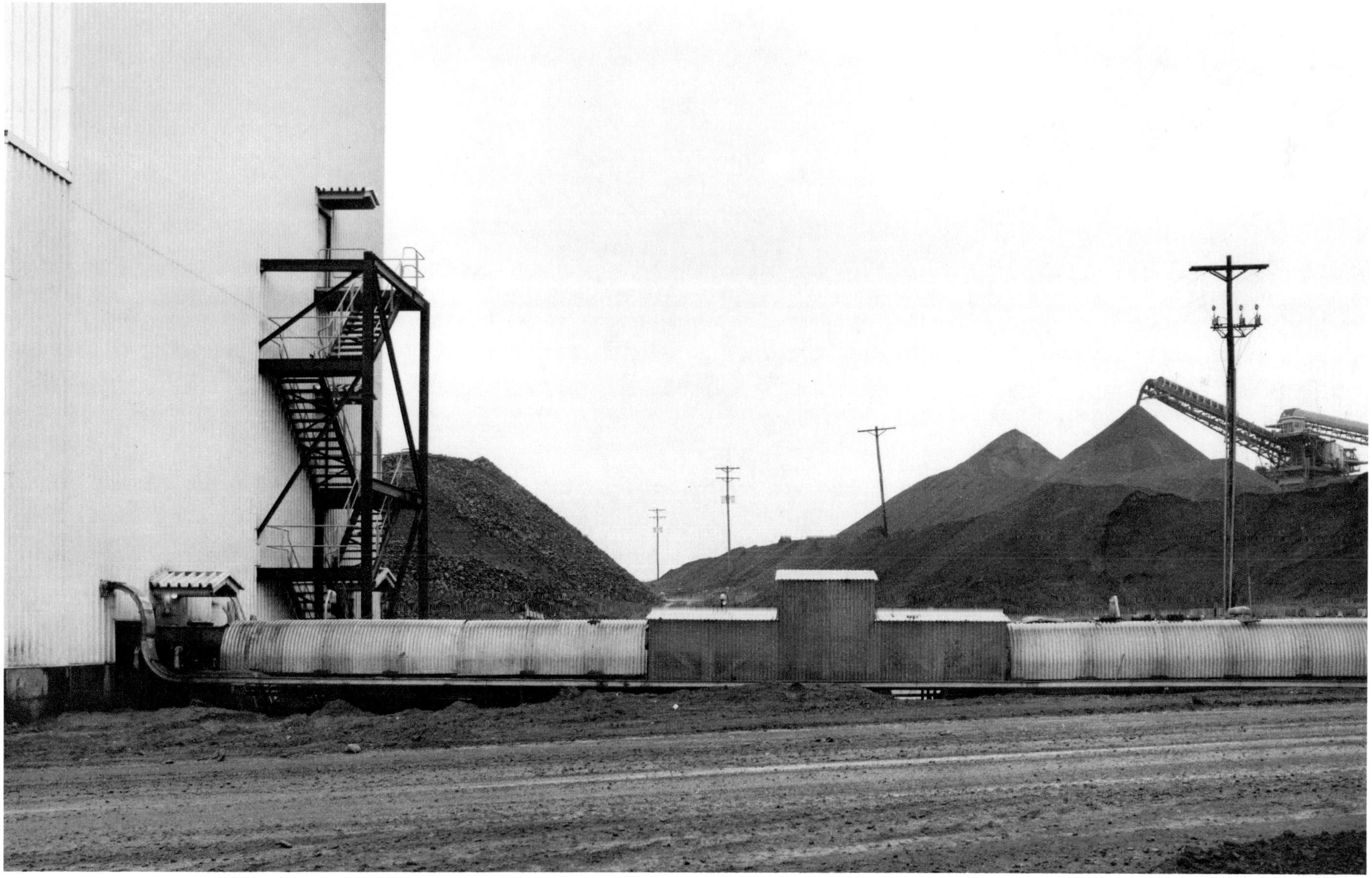

Primary autogenous (crushing) mills, taconite plant.

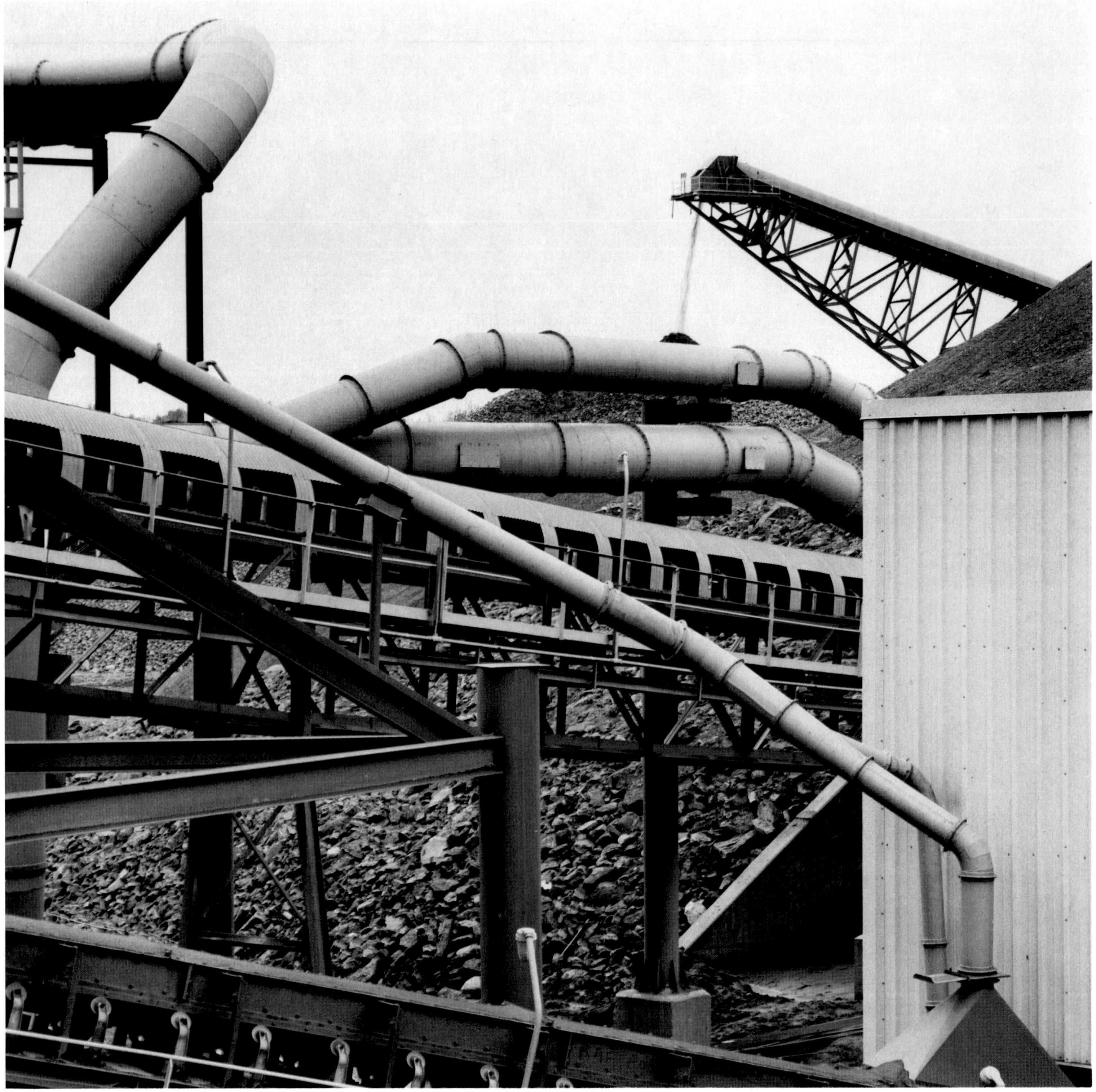

Taconite plant.

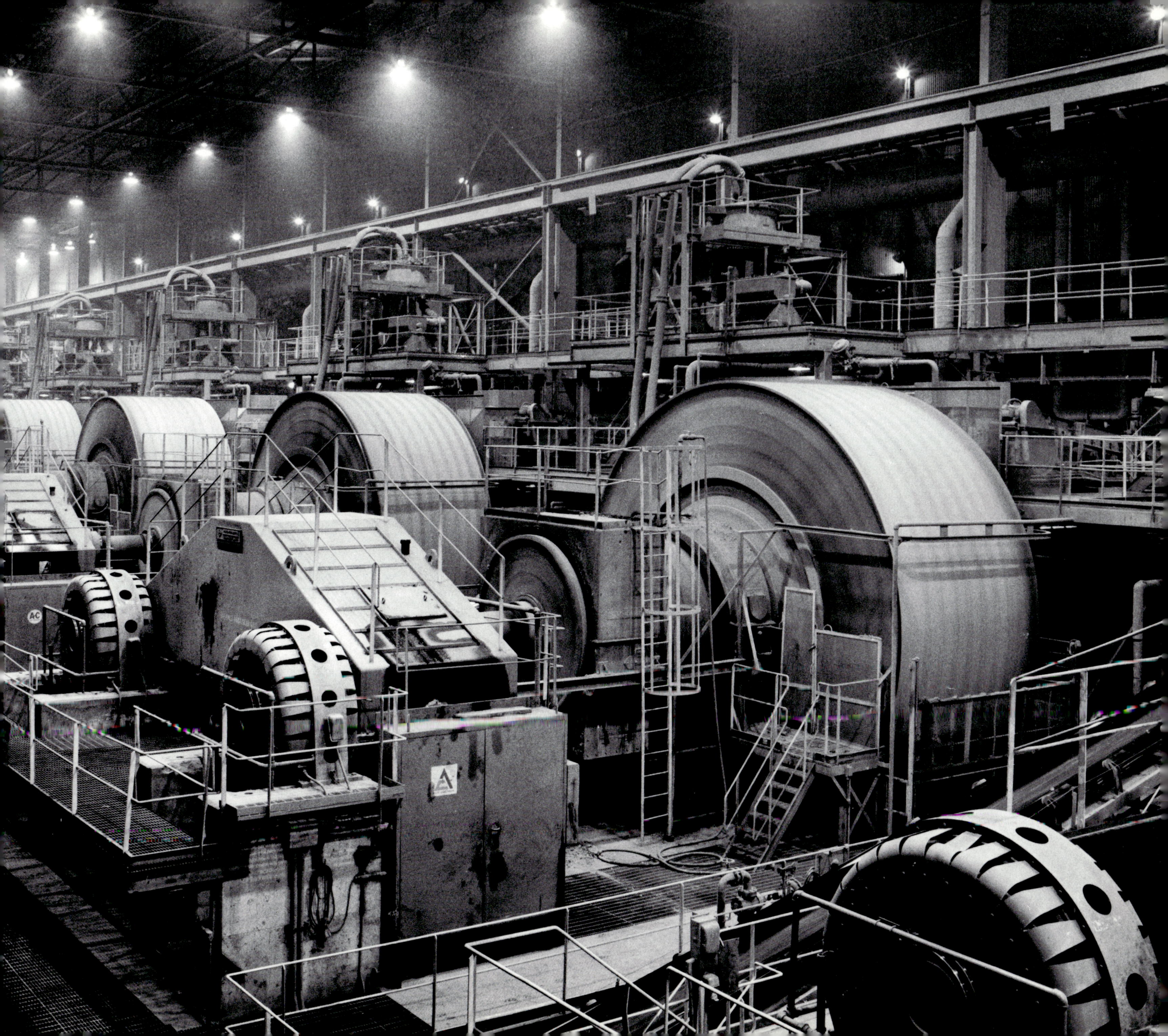

Taconite plant.

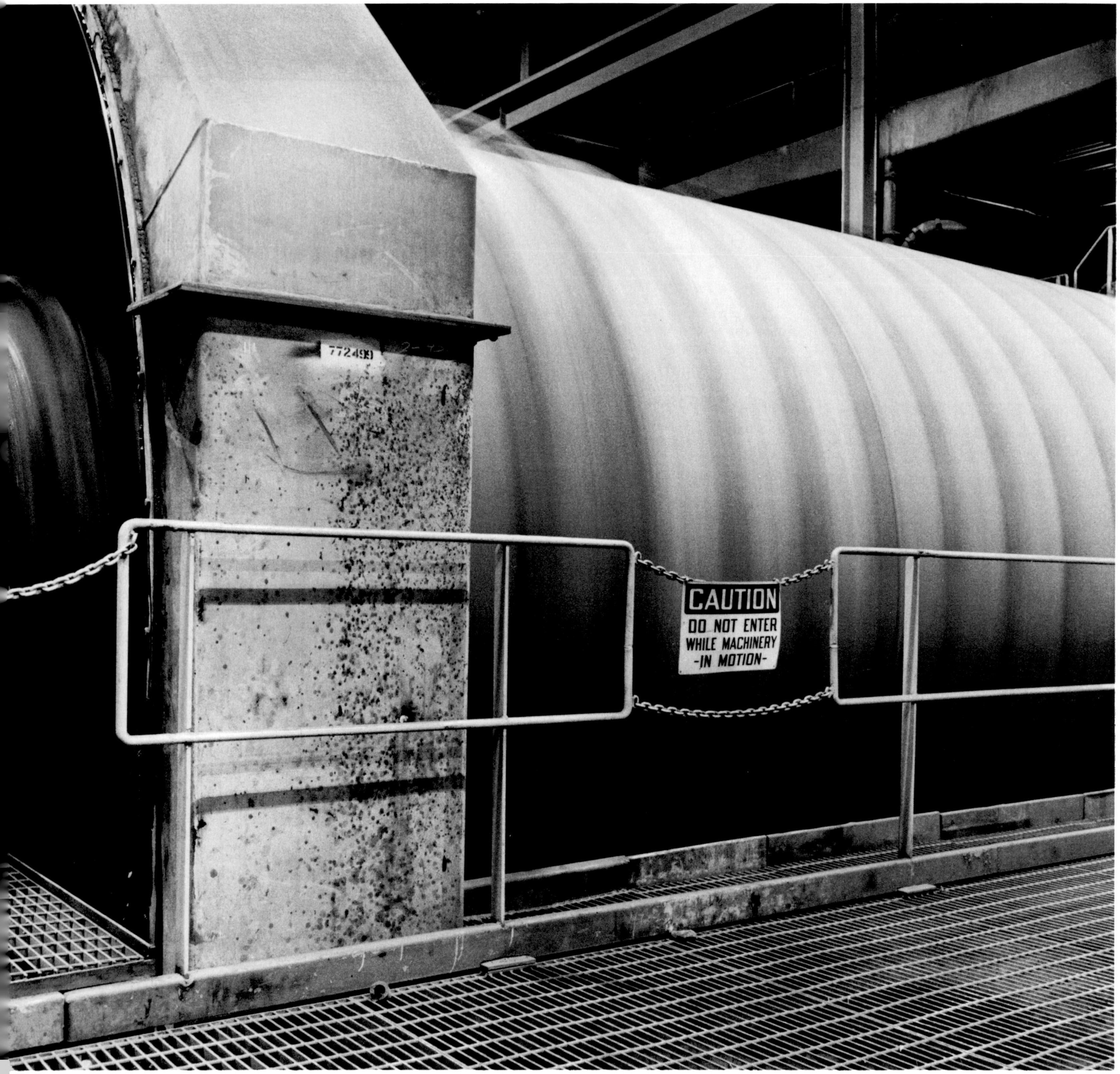

Pebble mill.

Magnetic separator.

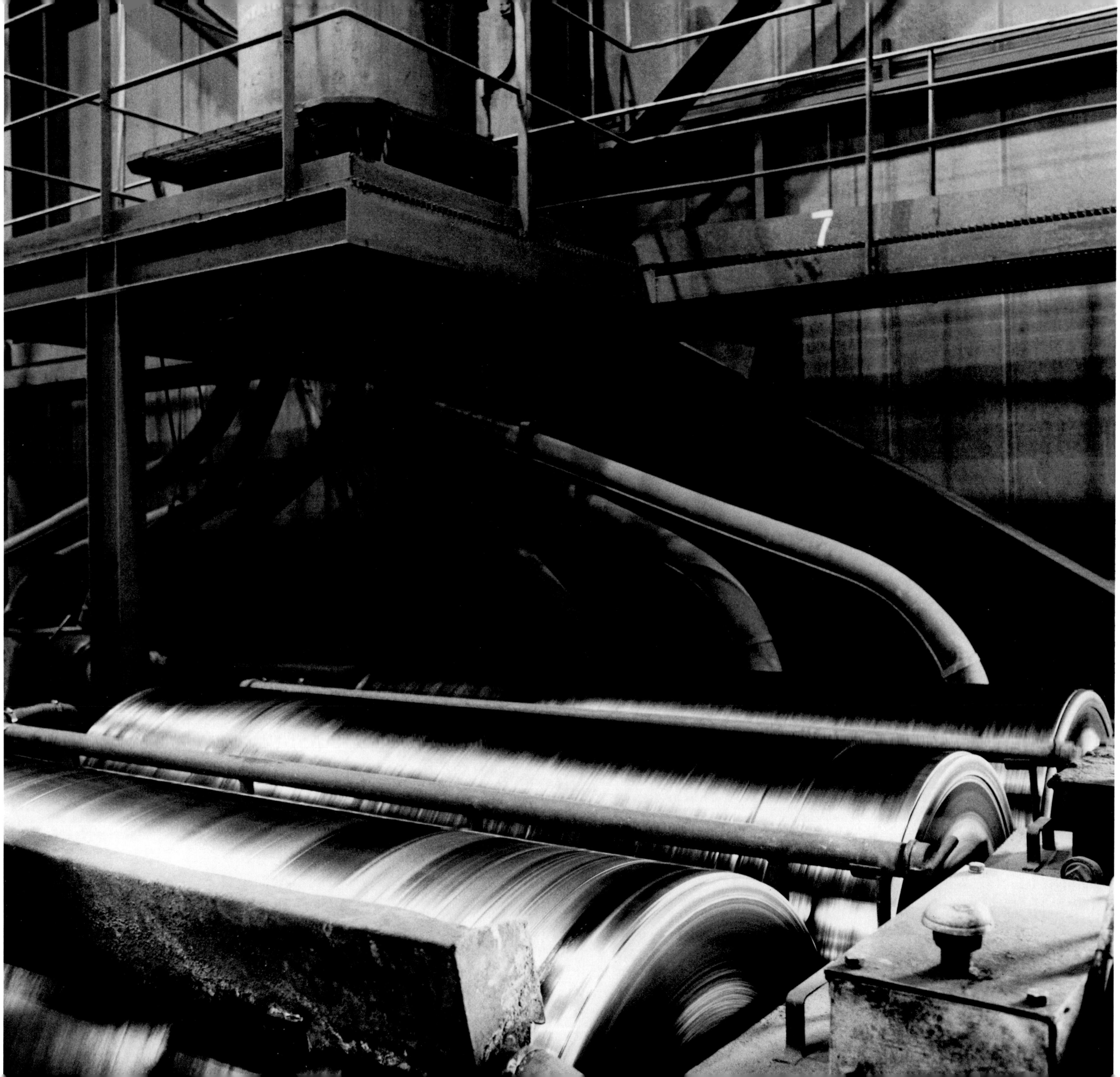
7

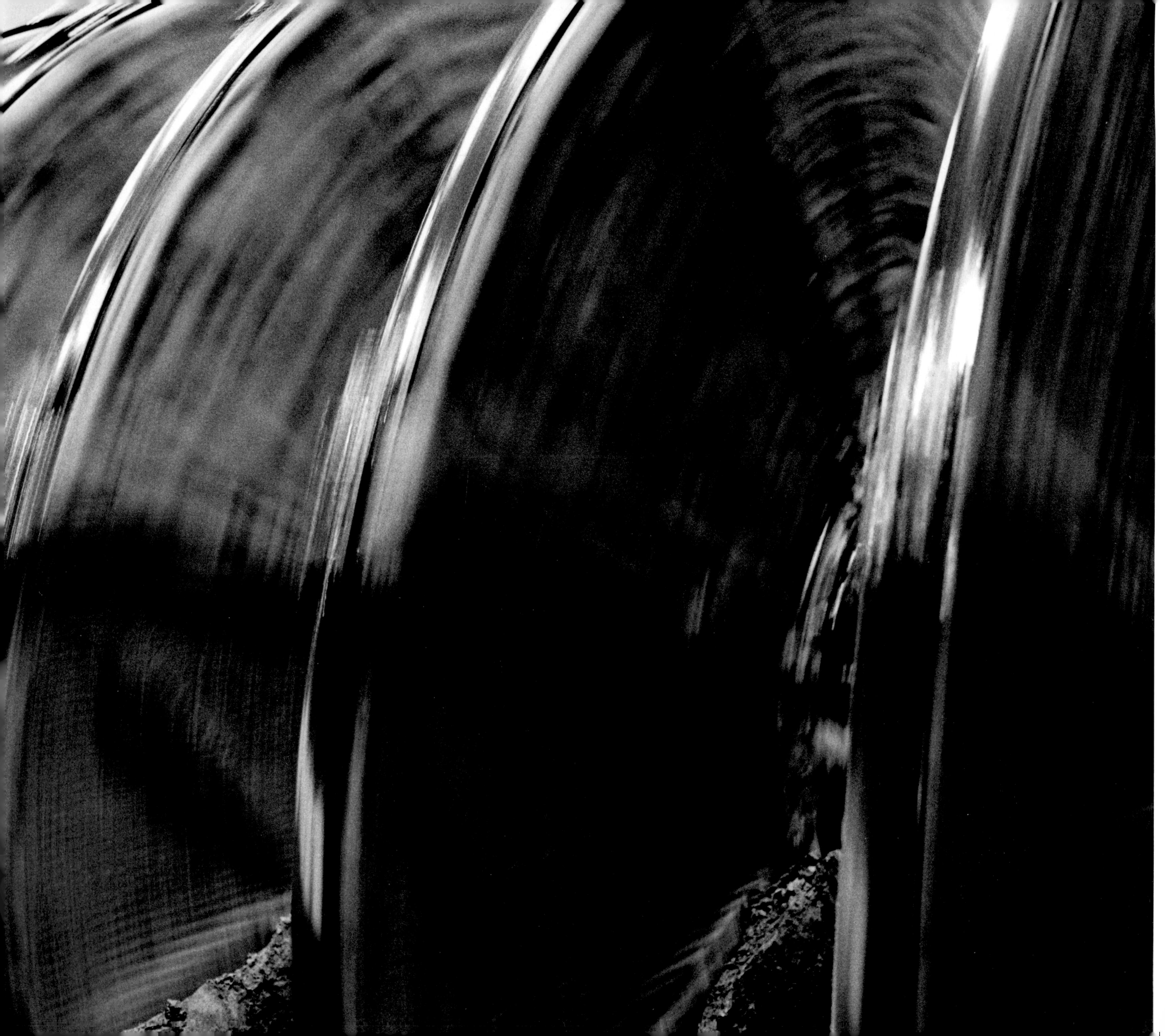

Taconite plant.

Overleaf: taconite pellets.

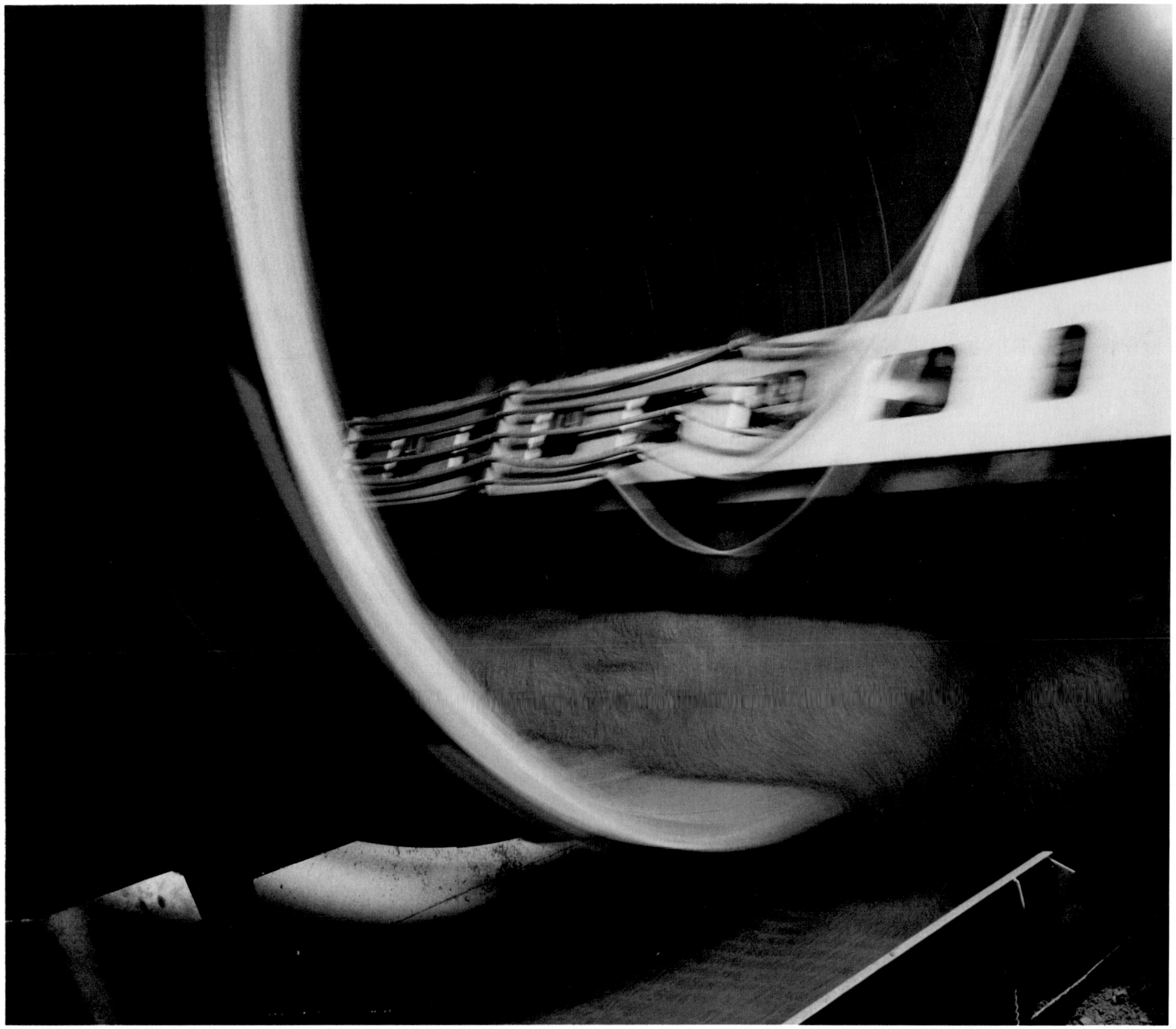

Dewatering filters.

Balling mill.

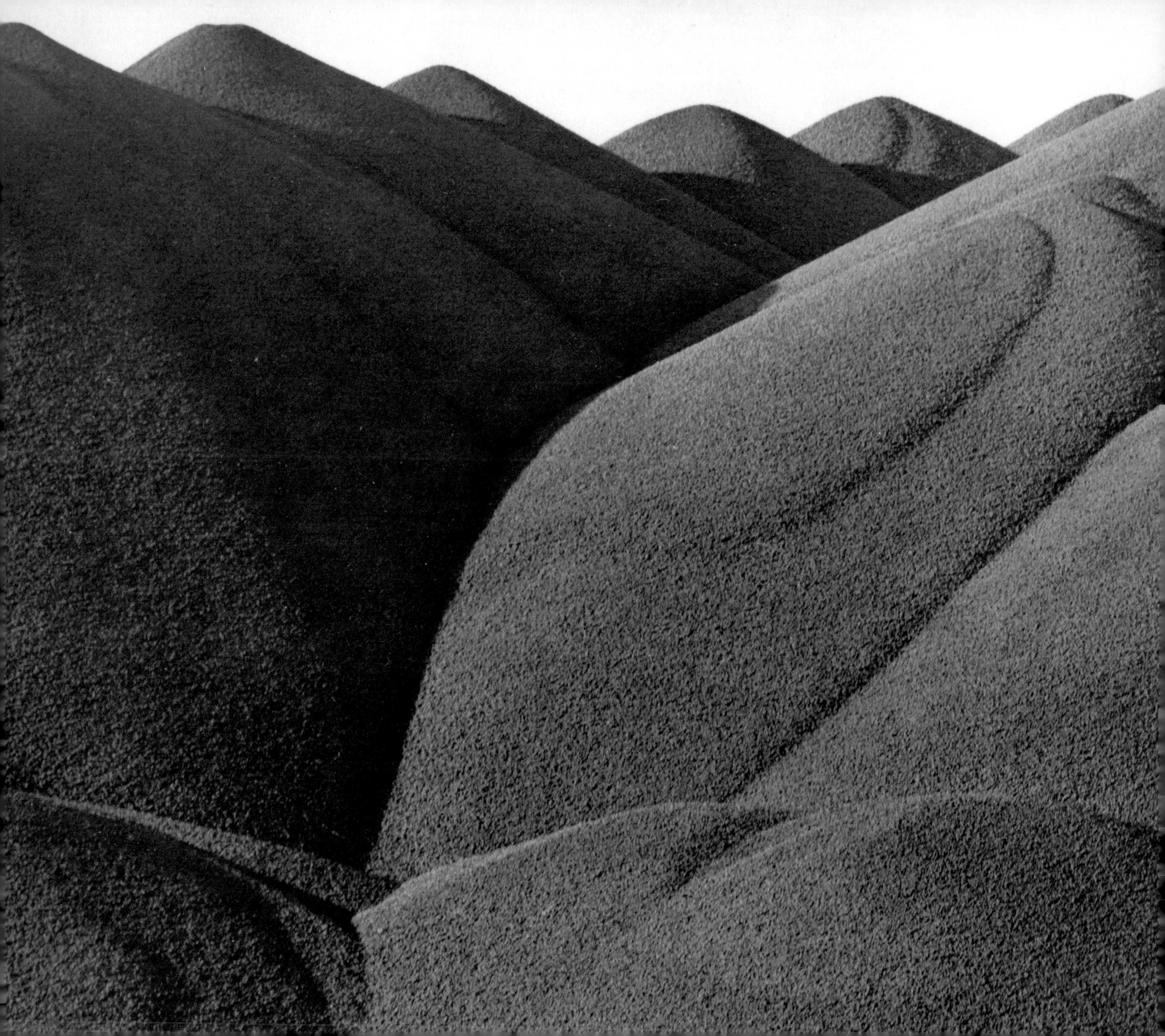

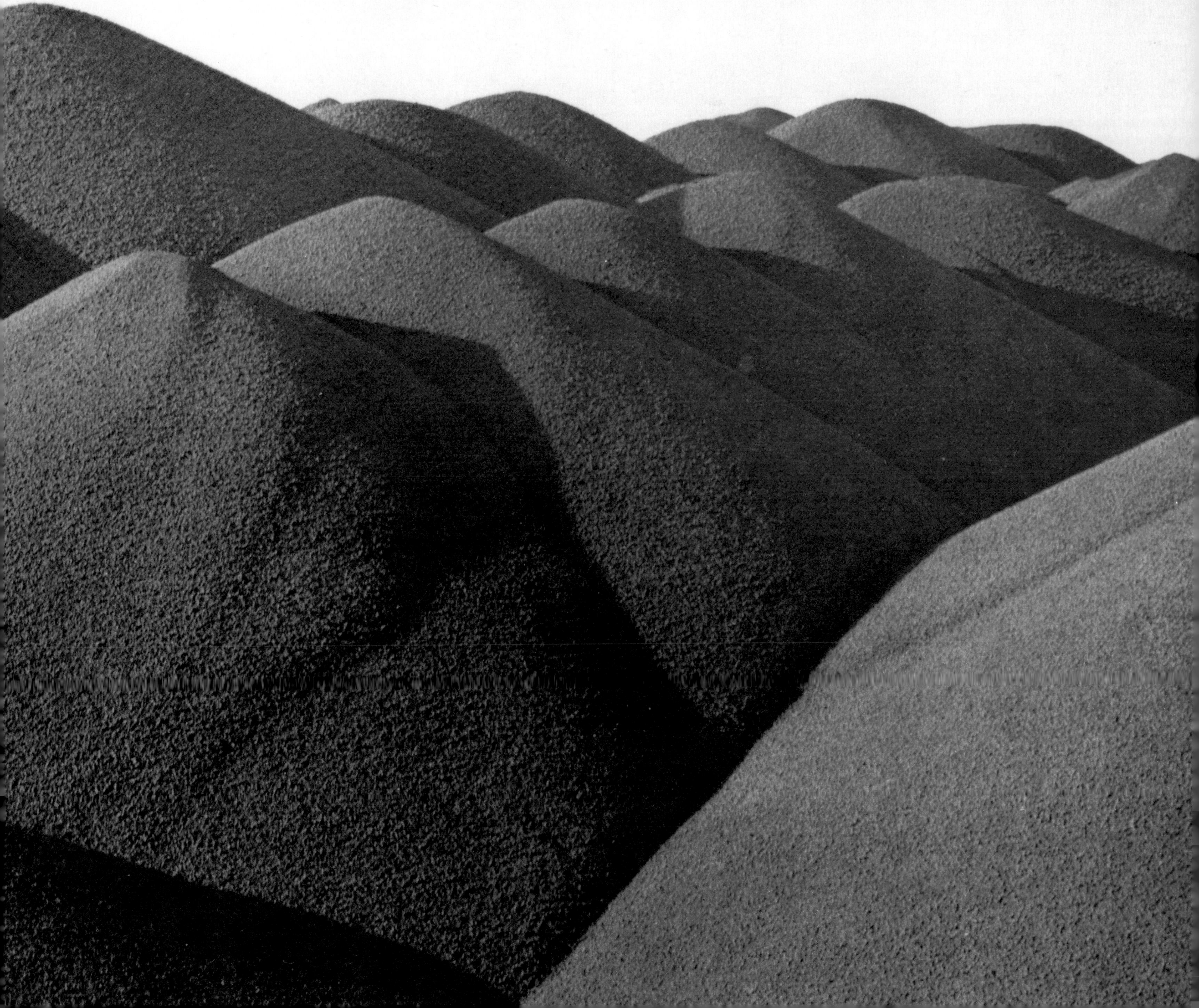

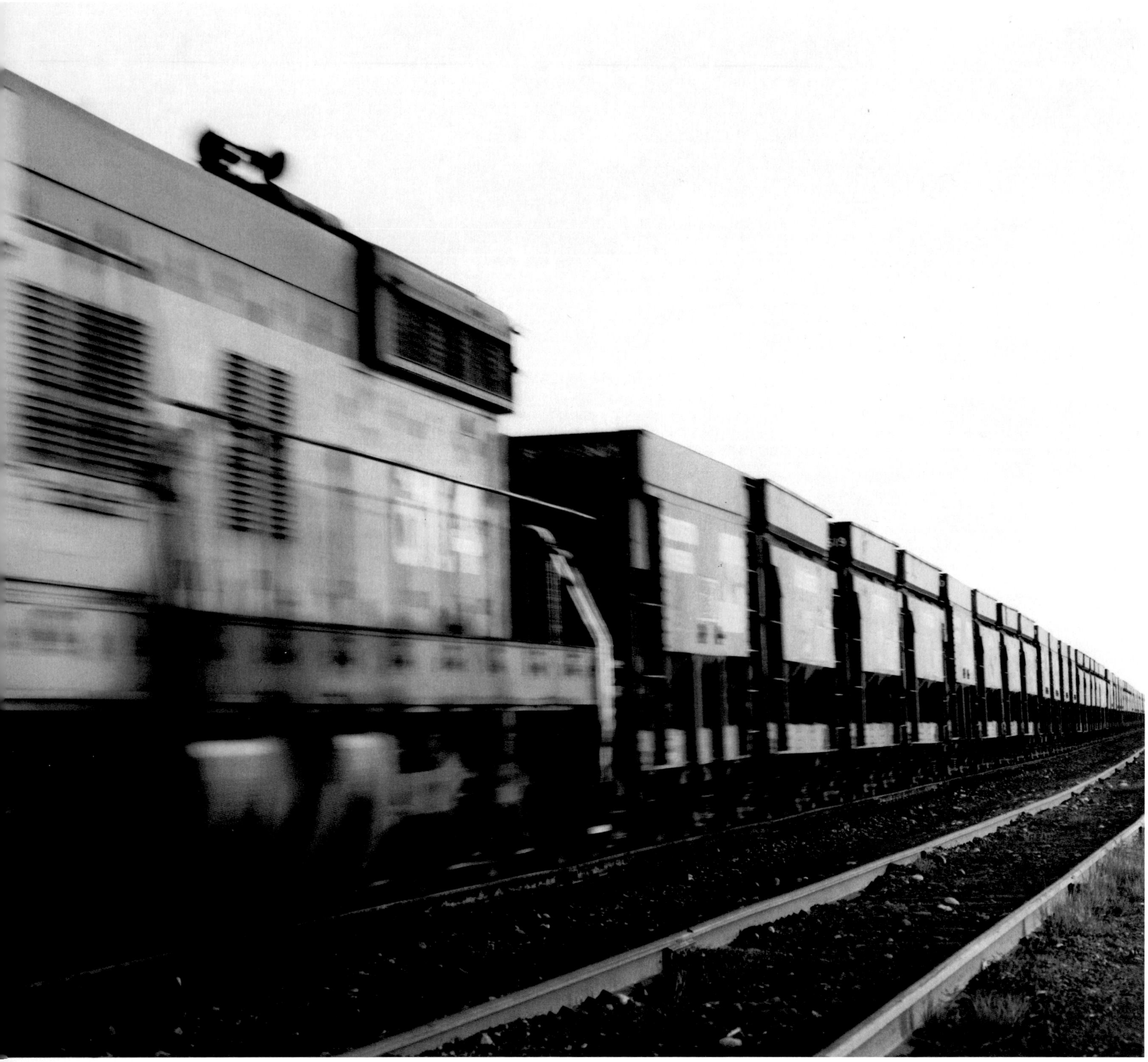

Ore train.

Taconite stockpile.

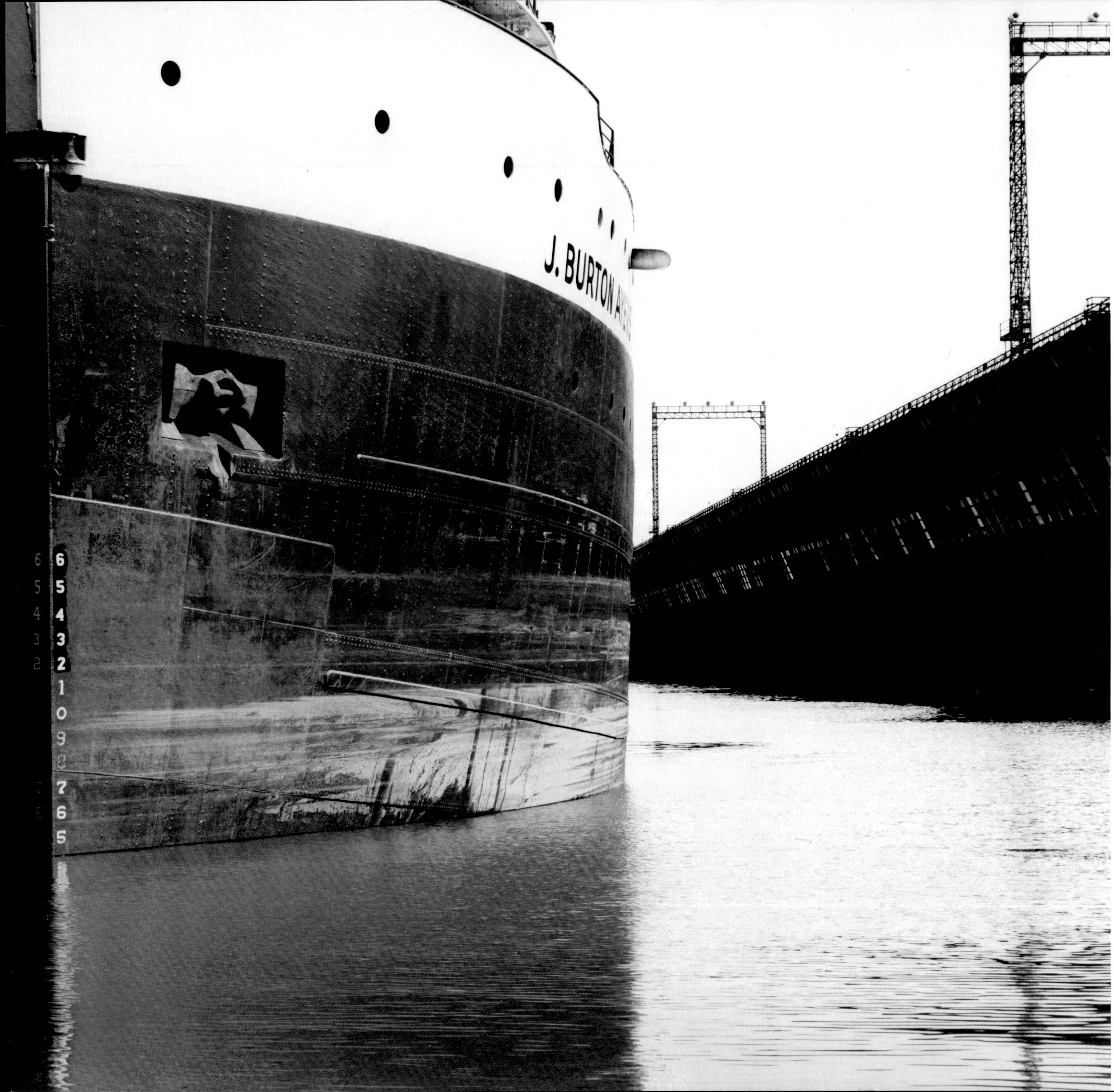
J. BURTON AYERS

Ore docks.

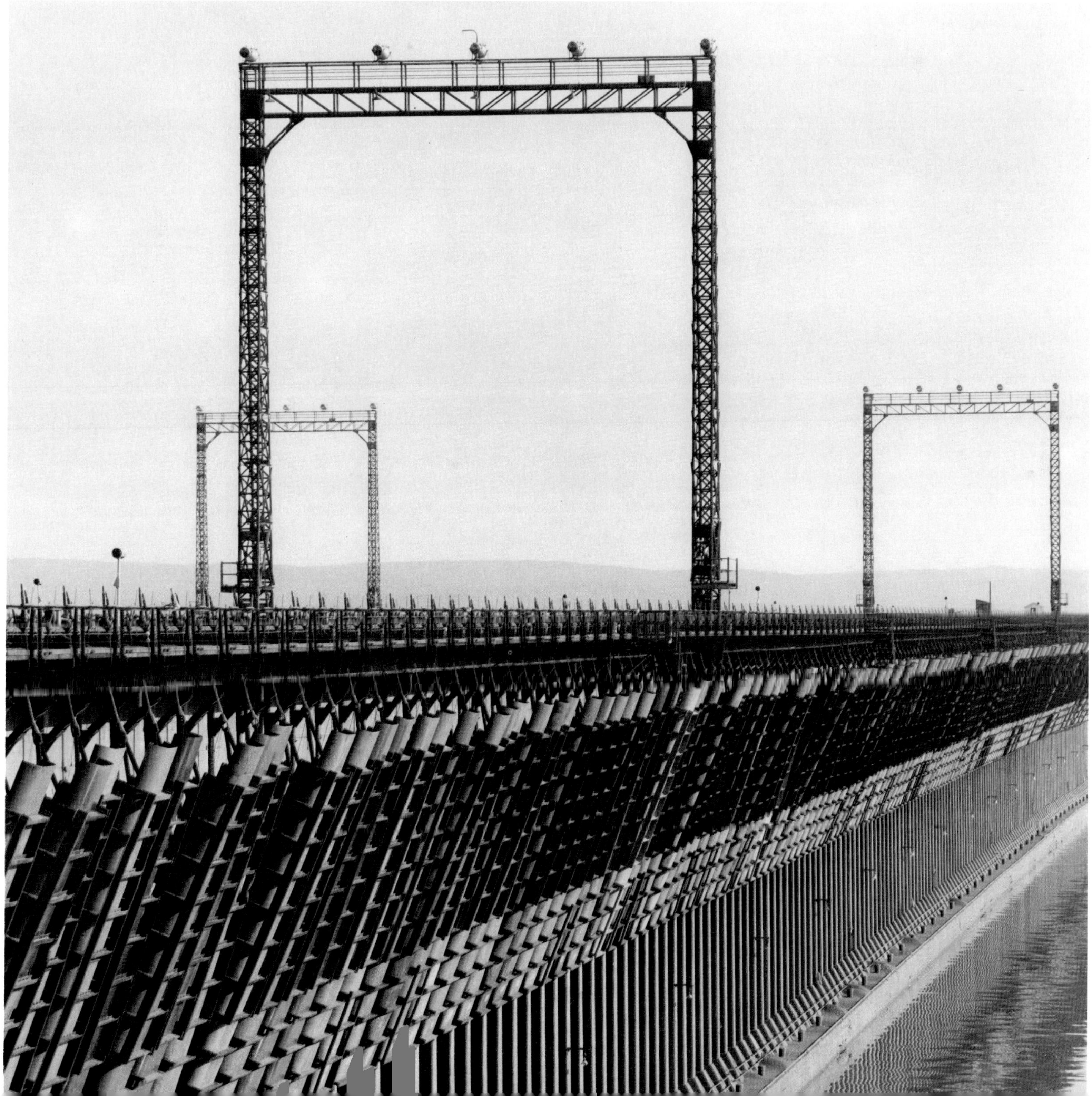

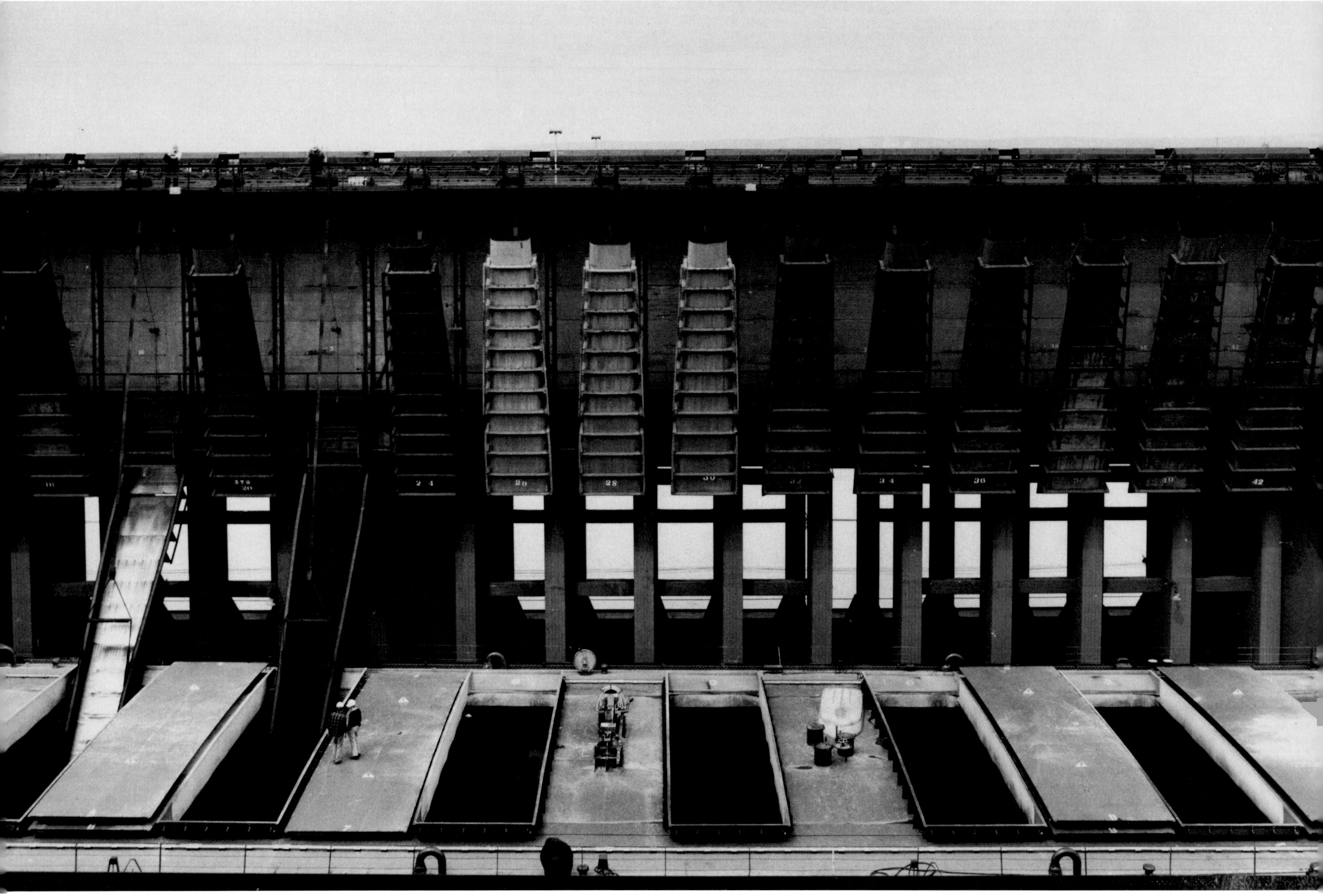
2 4
28
3 4
36
42

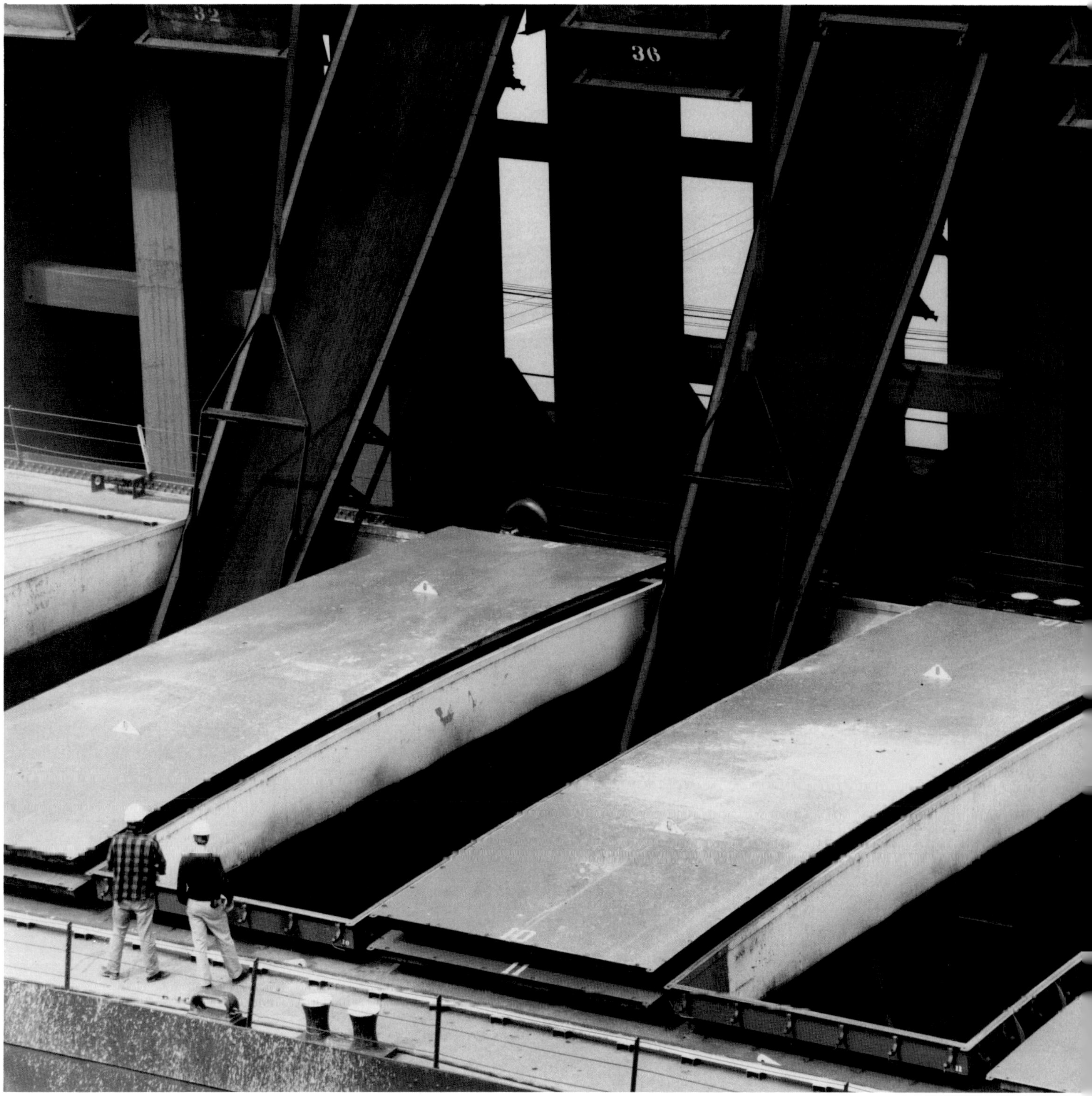

Loading taconite pellets into an ore boat.

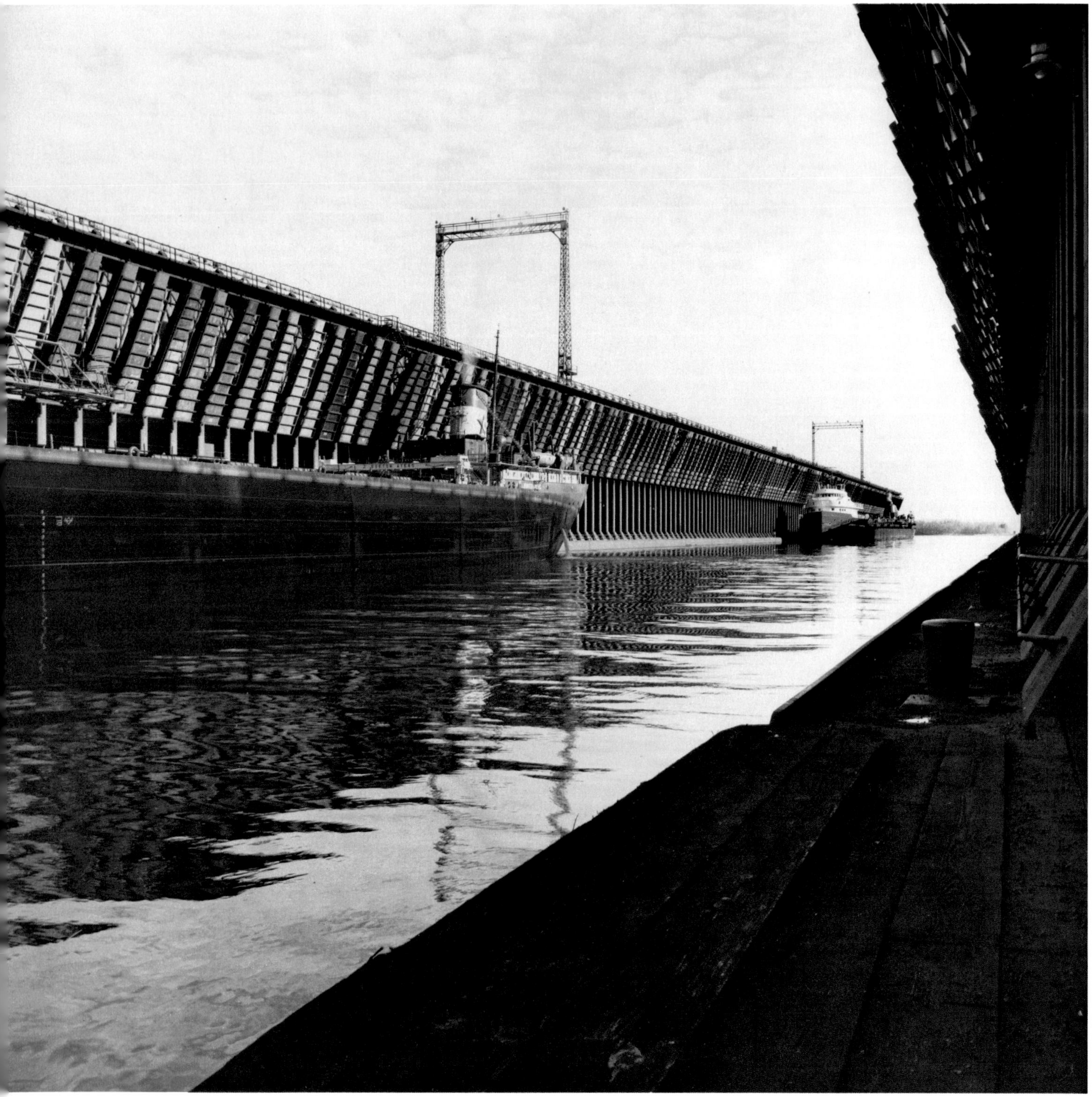

Ore docks.

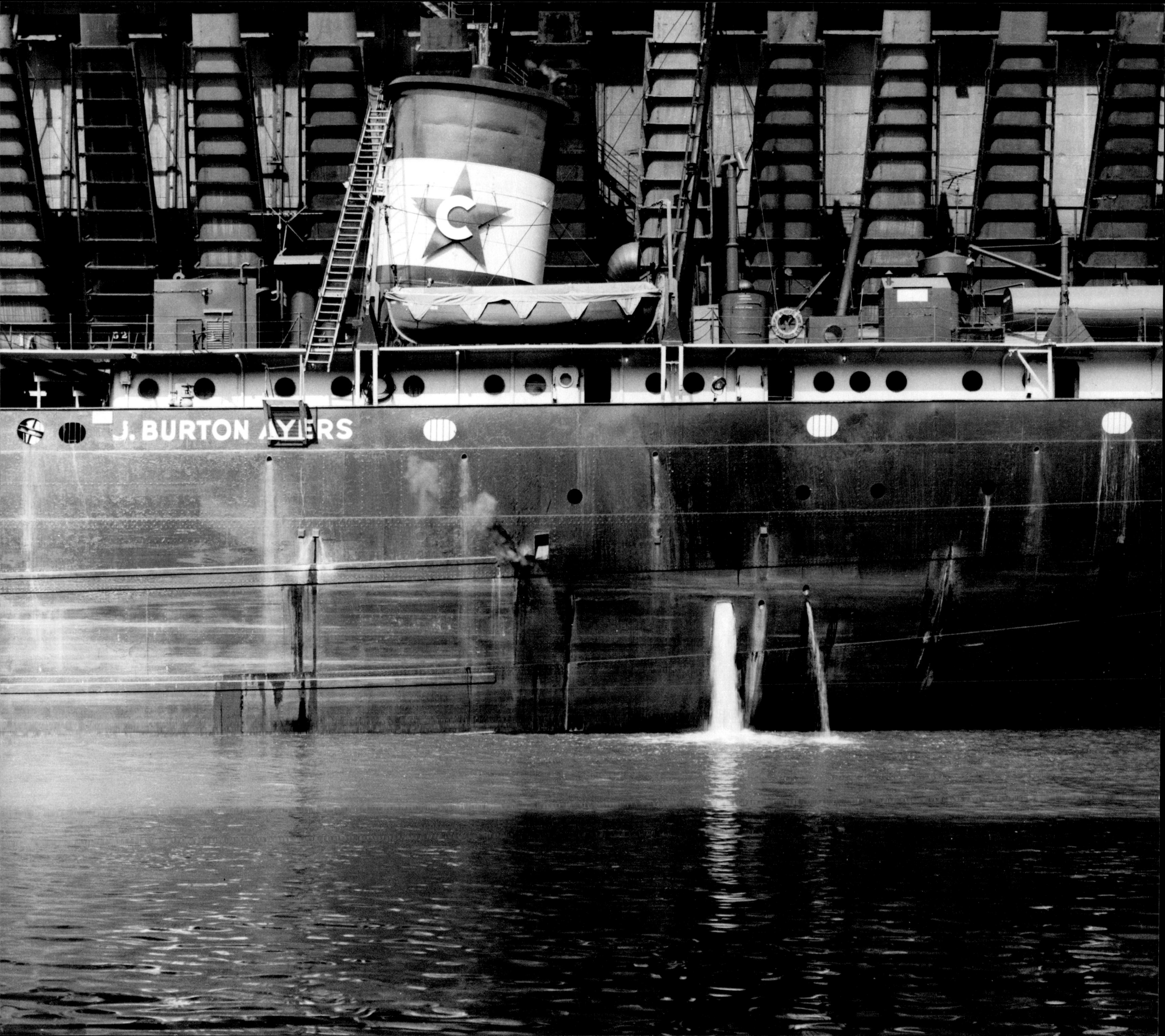
J. BURTON AYERS

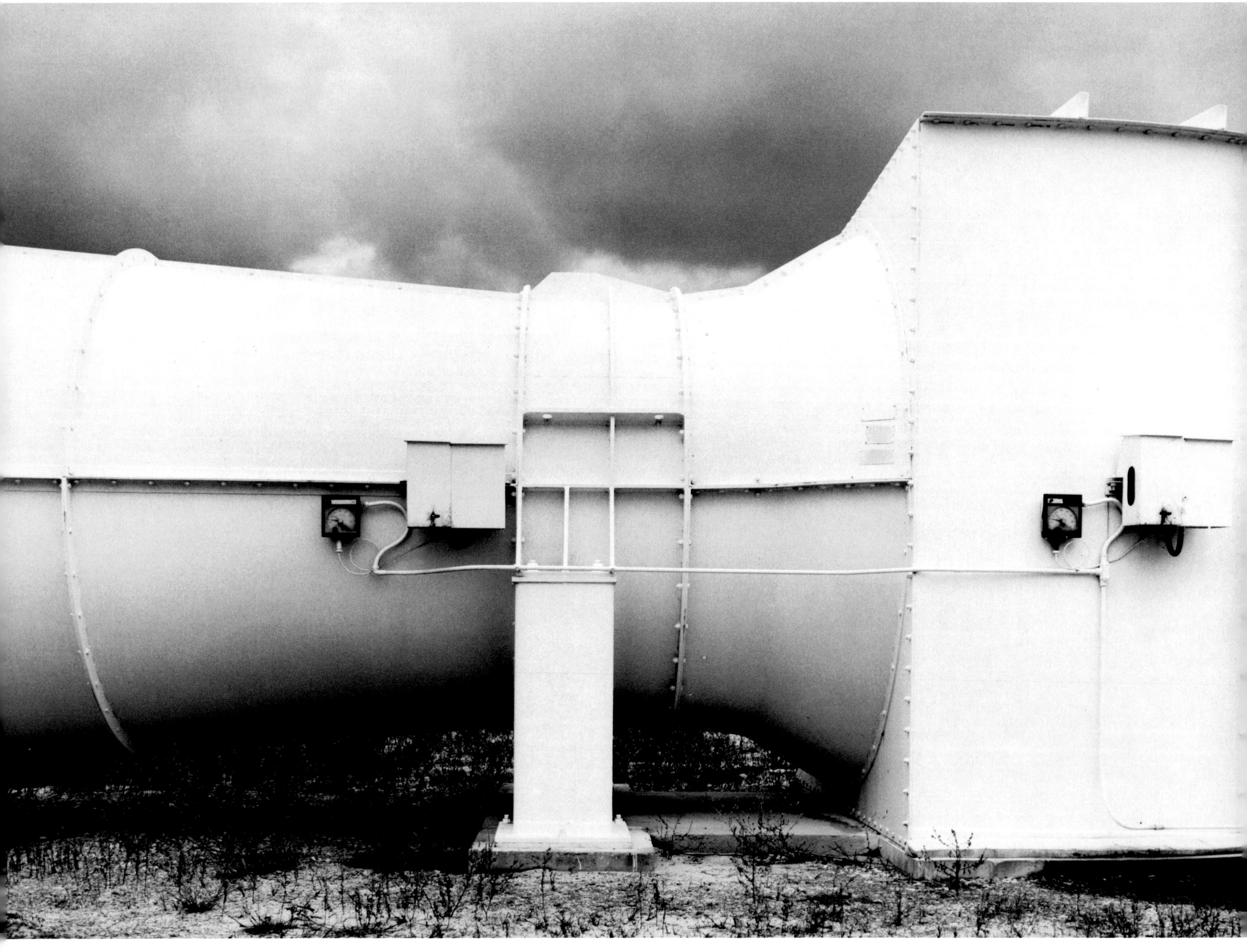

Raw Materials: Coal/Coke

Apart from ore, coal is the most important raw material used in making iron. In fact, more coal is consumed in the steel-making process than in any other operation—save, of course, the generating of electricity. Each year some forty million tons of coal are consumed in the form of coke by the blast furnaces of America.

Coke is the basic fuel of the blast furnace. Equally as important, it provides that crucial ingredient—carbon—whose presence is necessary in both iron and steel. And although it is derived from coal, coke is of a very different nature. Technically, it is "the coherent cellular residue resulting from the distillation at high temperatures of the volatile substances in coal." More simply, coke—which is primarily carbon—is what is left of coal after the gases and tars—hydrocarbons—have been removed. Unlike coal, coke is hard and porous and able to resist the crushing weight of the iron ore and other materials within the furnace. Coke burns uniformly both inside and out. This quality and the fact that coke does not contain the sticky tars of coal prevent it from fusing together and clogging up the furnace, as coal would.

The metamorphosis of coal to coke takes place inside long and narrow airtight chambers known as coke ovens. The ovens are usually arranged together in rows of eighty or more to form coke-oven batteries. The basic recipe for coke is universal: each oven is charged from the top with sixteen to twenty tons of pulverized coal, which bakes from twelve to eighteen hours at a temperature of approximately 2,400 degrees F. The entire procedure must take place in the absence of air; otherwise, the coal would burn. When ready, the material within the oven has coalesced. The doors at either end of the oven are opened, and the coke "brick" is pushed by a ram from one end into a "quench car" waiting at the other end. The coke is then carried to the quenching station, where it is doused under a spray of water, which simultaneously cools it and creates a most spectacular display of steam. The coke is generally then conveyed directly to the blast furnace for immediate use.

The nature of coke-making has earned the process the reputation of being the steel industry's dirtiest operation. In its infancy, coke-making was indeed a foul business because no effort was made to recover any of the materials that were driven off during baking. The result was not only a frightful degree of air pollution but the loss of some very valuable substances as well. In addition to reducing air pollution considerably, the sophisticated "by-product coke ovens" of today have also made it possible to reclaim a variety of substances during the coking process. Aside from large quantities of gas, which is used as fuel to heat the walls of the ovens themselves, a wide spectrum of chemicals ranging from ammonia to xynol are now recovered. Many of our synthetic fibers, printing inks, and aniline dyes are obtained from these chemicals, along with the bases for such diverse items as aspirin and perfume.

Coal is our most plentiful fossil fuel, but not all coal can be made into coke. Anthracite and lignite, the hardest and the softest coals respectively, are not eligible. Coking coal must be bituminous, and it must have a very low ash content. Additionally, sulfur and phosphorus, elements whose presence in steel is deleterious, must be present only in very small quantities. These qualifications limit the types of coal that can be classified as metallurgical-grade coking

Ventilator fan at a coal mine.

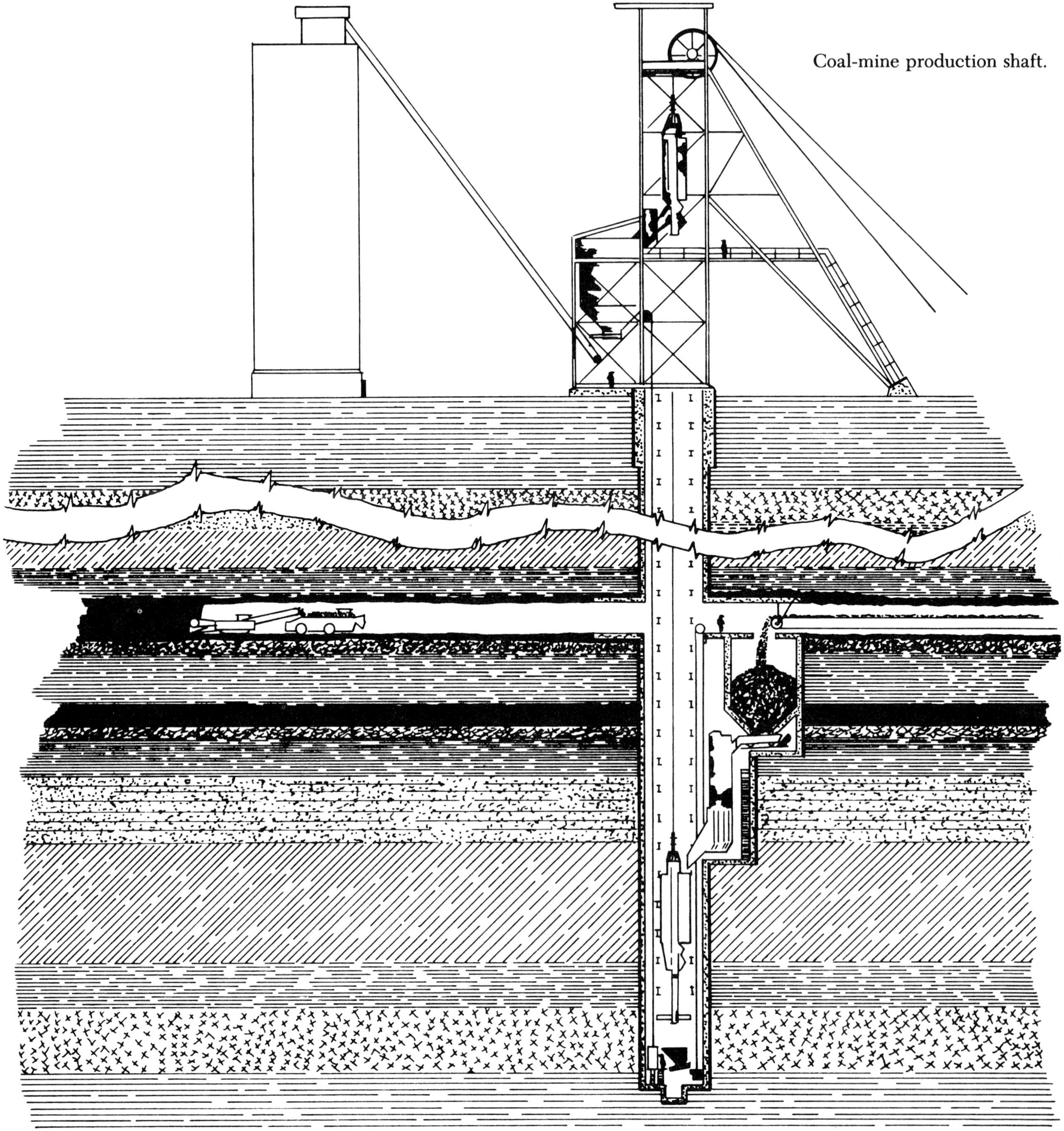

Coal-mine production shaft.

coal, or "met-coal," and thus be considered suitable for use in making iron.

The United States happens to boast the lion's share of the world's supply of good coking coal. Eighty percent of this has been mined in West Virginia, Pennsylvania, Kentucky, and Alabama—a geographical distribution that in turn has been one of the decisive factors in the location of our steel-making centers. But just as the high-grade ores of the Lake Superior region proved to be finite, so too are the reserves of high-grade coking coals. Steelmen have had to find ways of utilizing coal with higher amounts of sulfur and phosphorus than were once considered acceptable. They have dealt with the problem in much the same way that they have coped with the task of turning taconite into an ore.

All coal destined for the coke ovens must be crushed, washed, screened, and cleaned in a variety of ways after it is mined to remove the unwanted elements, primarily ash and sulfur. Coal with a very high sulfur content must be subjected to an even more intensive purification process. At the present rate of consumption, our coal will last for many hundreds of years. However, in order to keep pace with the demand for coal of higher quality, the industry has had to modernize its operations. Coal-mining today, although hardly a kid-glove operation, is a far cry from what it once was.

Although at the present time no major American steel company uses coal obtained by strip-mining, strip-mining has become increasingly prevalent, especially in the midwestern and far-western states where our greatest reserves lie. From the point of view of the consumer and the miner alike, strip-mining is a boon. It is a relatively inexpensive and easy way to obtain vast quantities of coal. The procedure is virtually the same as the open-pit method of mining ore: the overburden is stripped away to expose a coal seam; once this is laid bare, power shovels can easily remove many thousands of tons of coal a day. Strip-mining, however, leaves long scars on the land—scars that circumcise the tops of mountains, scars that will remain unhealed long after the men that made them and those who have used the coal are gone. Much of the overburden also happens to be some of our richest farmland and most beautiful countryside.

In 1977 legislation was enacted requiring mine owners to "reclaim" all the land that has been mined—to return it as nearly as possible to its "original state." A tall order, this is a major step in the right direction, although its success of course depends upon whether or not the mine owners themselves comply.

At the present time, coking coal is still obtained by underground mining methods, but these methods are a far cry from the "pick and shovel" days of mining. Today machinery is replacing man in the mines. Currently more than 65 percent of the coal extracted from beneath the surface of the United States is obtained by continuous mining methods: huge mechanical moles—continuous mining machines—rip and claw at the face of the coal seam. A single man at the controls of one of these machines can mine as much as twelve tons of coal a minute. The coal is then carried on conveyor belts and rail cars from the depths of the mine to the surface or to a shaft where it will be hoisted aboveground in skip cars. When the coal reaches the sur-

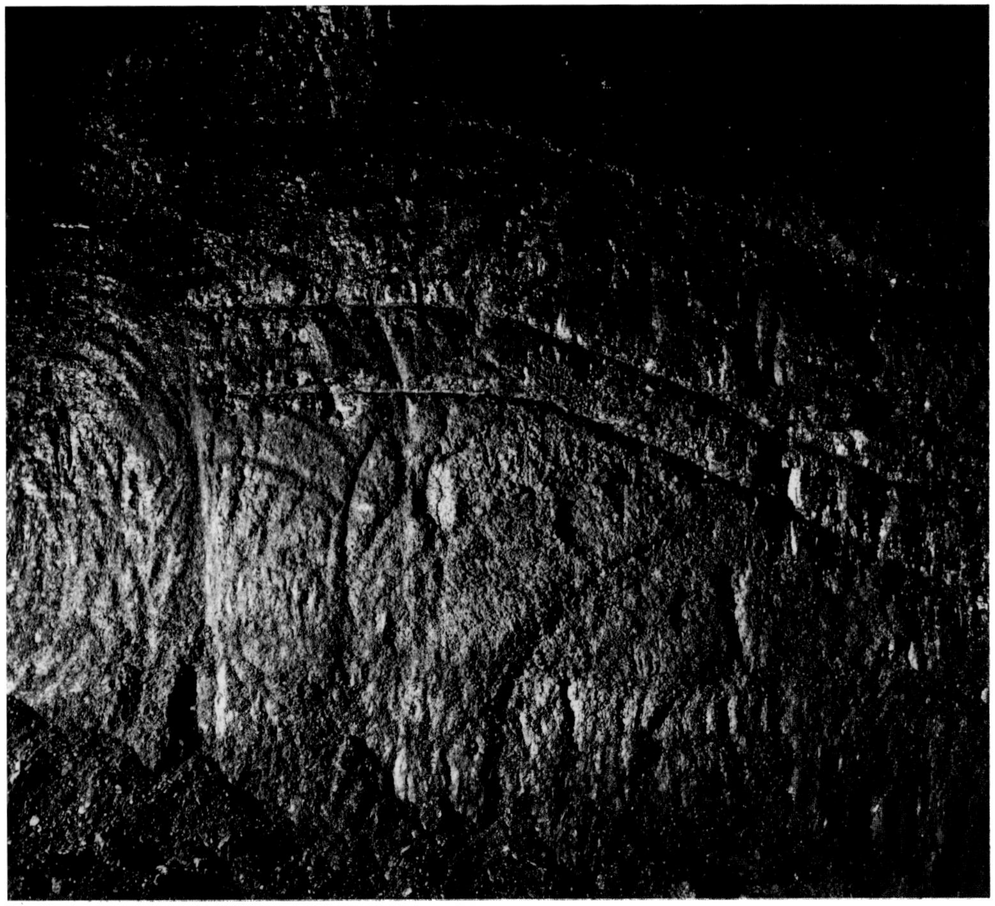

face, it will be processed for shipment to the coking plants.

Whether they utilize mechanized methods or not, underground mines are organized along the same lines. Each mine has a minimum of two access passages—shafts—that are driven from the surface of the land to the coal bed. One shaft has an elevator—the "cage"—to transport men and machines; the other—the "production shaft"—is used to haul the coal out of the mine.

Underground mines are of three types: shaft mines, drift mines, and slope mines. Shaft mines are employed when the coal lies at great depths. In such situations, the access shafts are driven straight down vertically from the surface to the coal beds. Drift mines, on the other hand, are used to extract coal from seams embedded within hills or mountains. Here the shafts are driven horizontally along the seam from its outcrop on the mountainside. In the case of slope mines, as the title implies, the shafts either slant downward into a hillside to the seam or actually follow the seam downward from the surface.

The most widely used method of underground coal-mining in America is the "room and pillar" system. In this method, rectangular blocks of coal from forty to eighty feet wide are left in place at regular intervals to support the overburden above the space that is being worked rather than having all the coal removed from an area that is being mined. Once all the coal between the pillars of coal has been mined, the pillars are removed "in retreat," allowing the roof to collapse. This process must be executed with great precision so that the roof does not collapse while work is still in progress. In addition to the pillars, "roof bolts," sometimes six feet in length, are inserted into the ceiling above the working areas in order to bind the layers of overburden together and reduce the danger of cave-in.

In spite of all the procedural precautions, underground coal-mining retains an element of danger: cave-ins, explosions, exposure to poisonous gases. Even the specter of black-lung disease, although diminished, still haunts every coal miner.

Because of these potential hazards, coal mines have been subjected to safety regulations of unparalleled stringency by the federal government, the individual states, the mining companies, and labor organizations. As a result, mining is much safer today than ever before.

The experience of being hundreds of feet belowground in a place of perpetual darkness—relieved only by the probing beam of the miner's cap light—is like nothing else. Here is one of the few places on earth where for an ephemeral second a man is confronted with the virgin fabric of the earth. Until this moment that earth had lain protected and undisturbed for several hundred million years. Man has finally found his way to the hidden lair of primeval energy and, in an instant, consumes it.

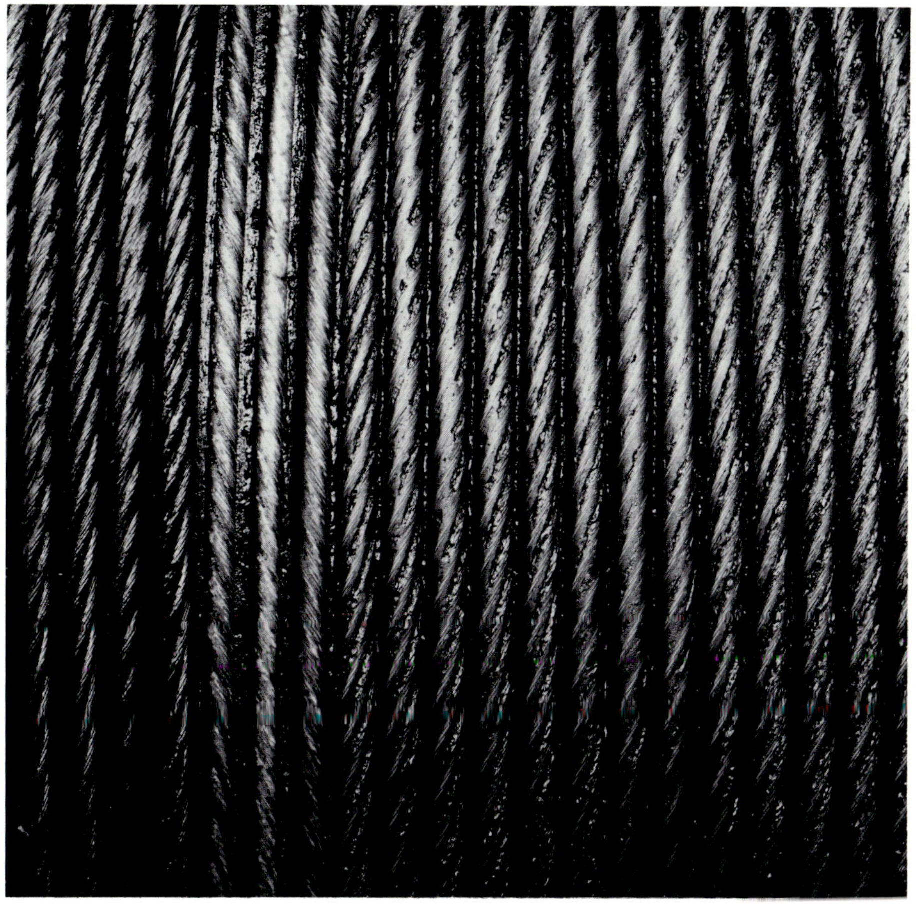

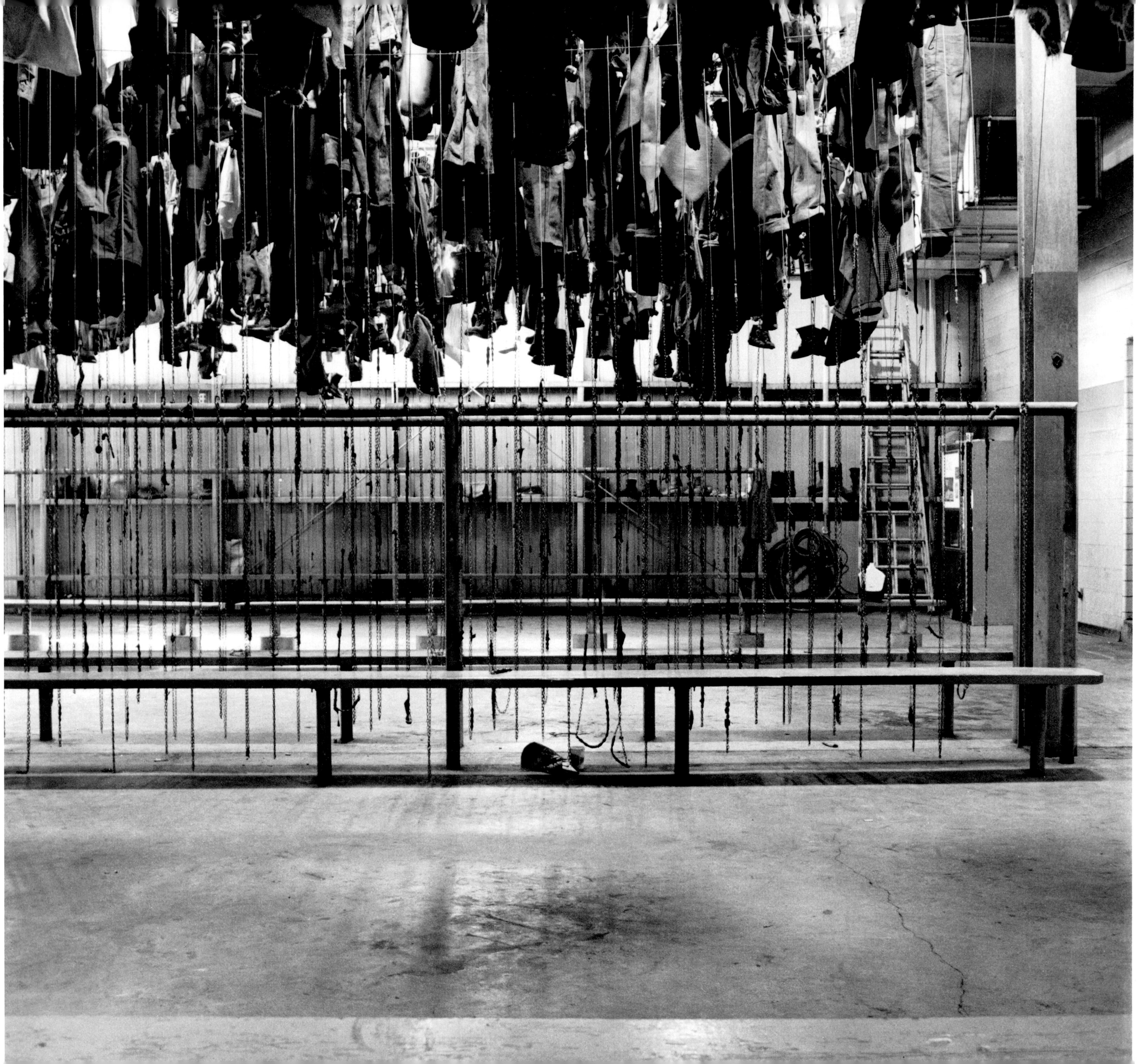

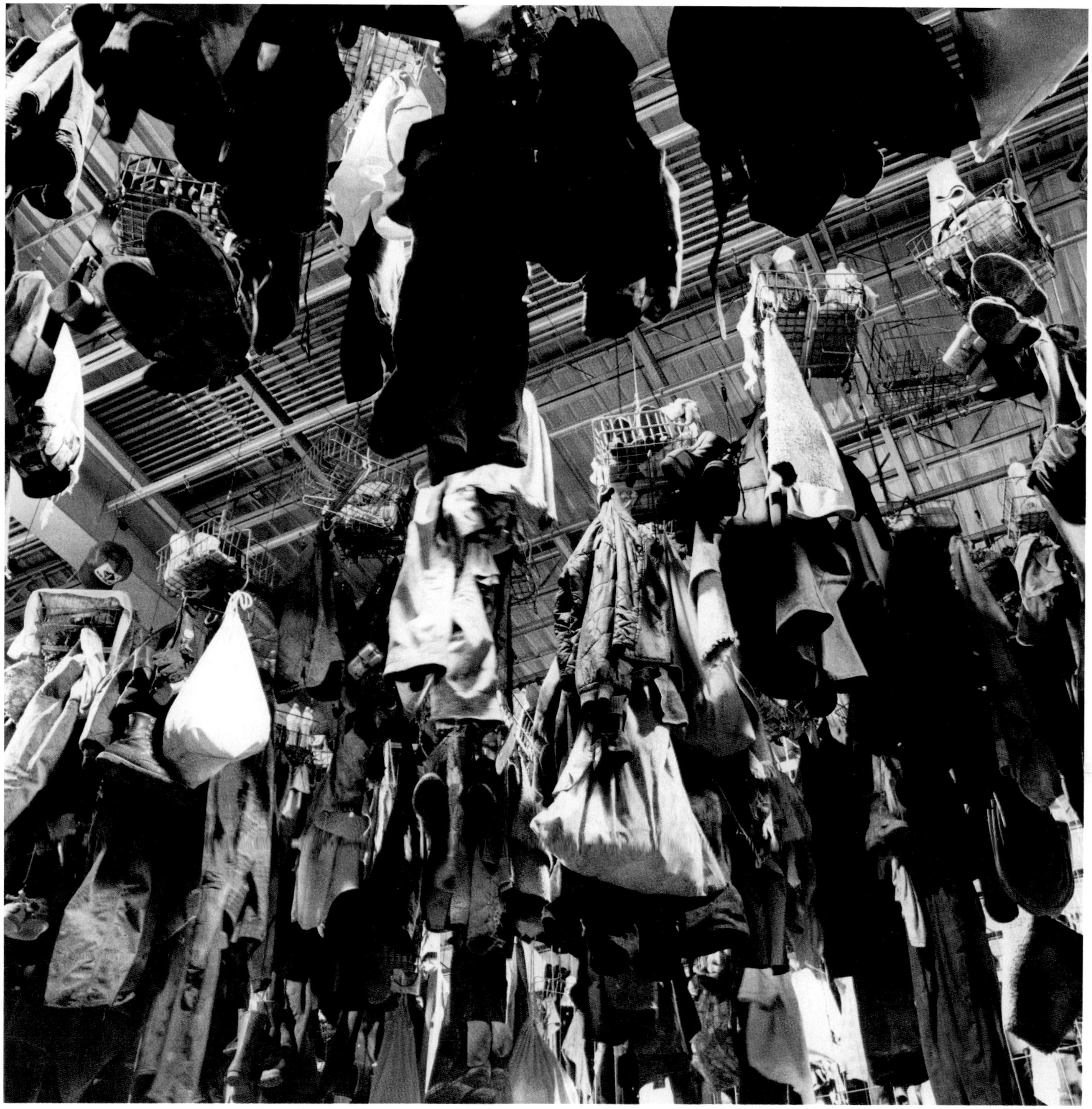

Washhouse at a coal mine.

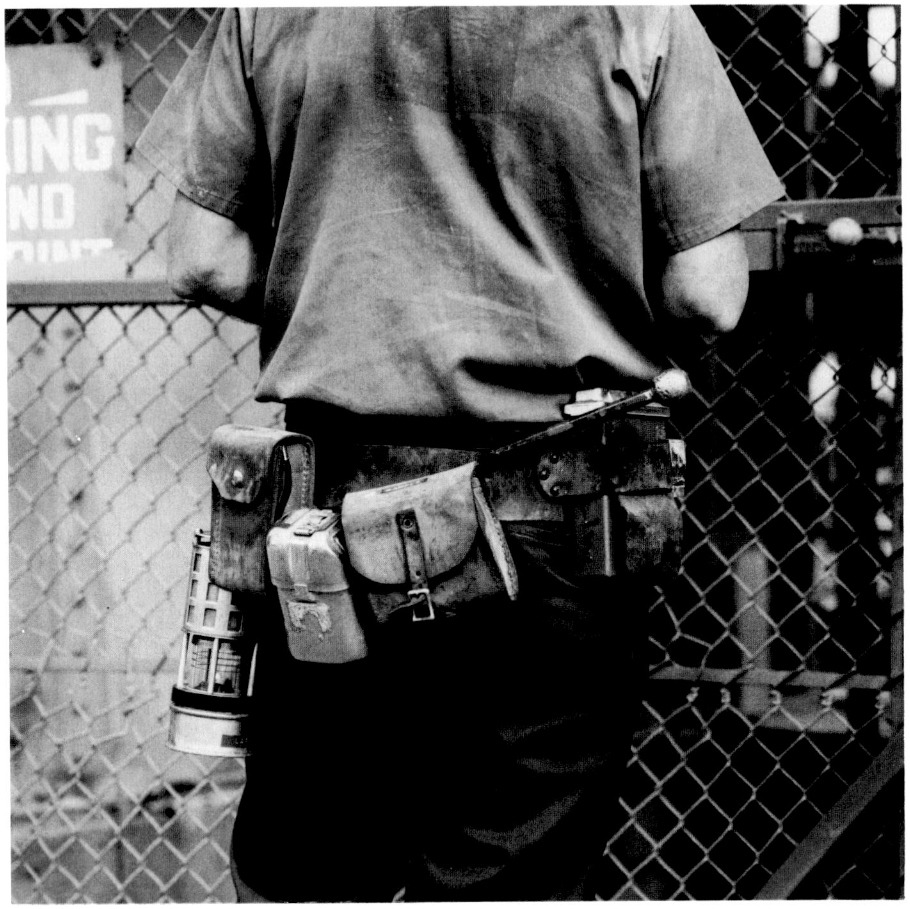

Coal-miner gear.

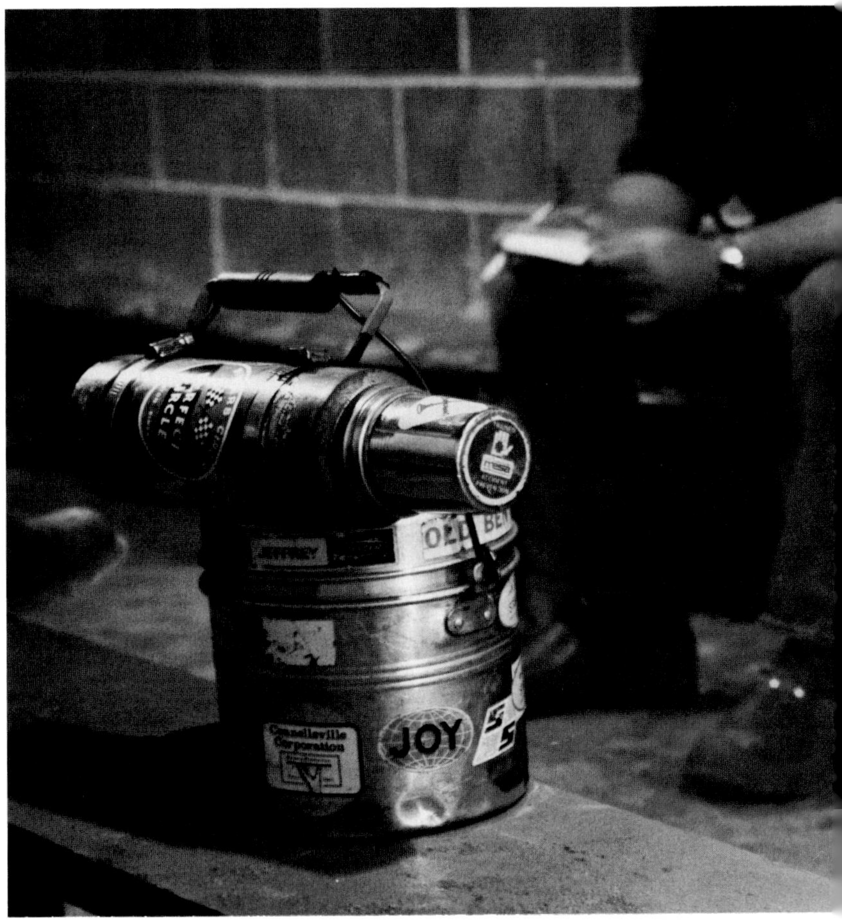

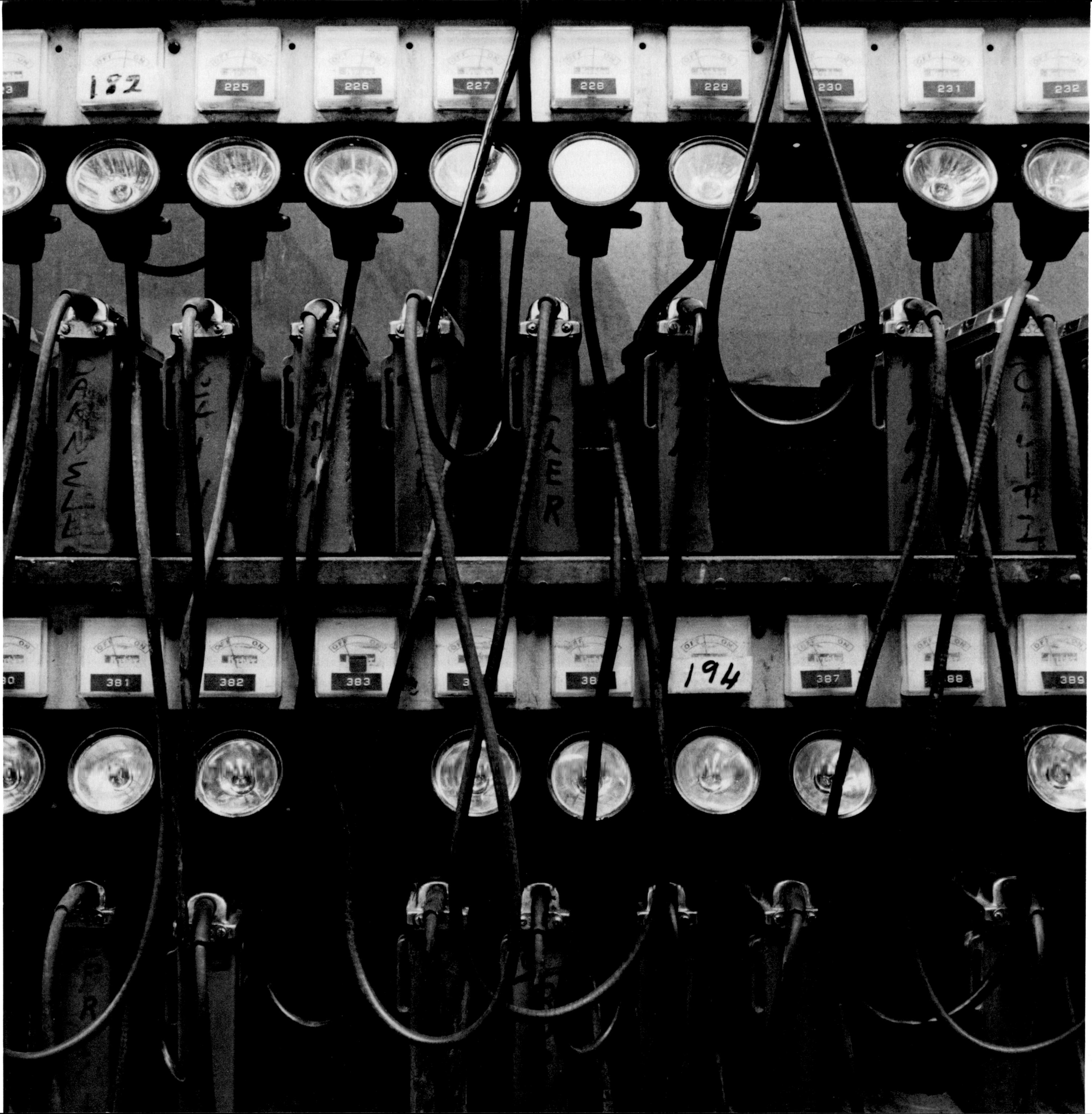
182
225
226
227
228
229
230
231
232
381
382
383
194
387
389

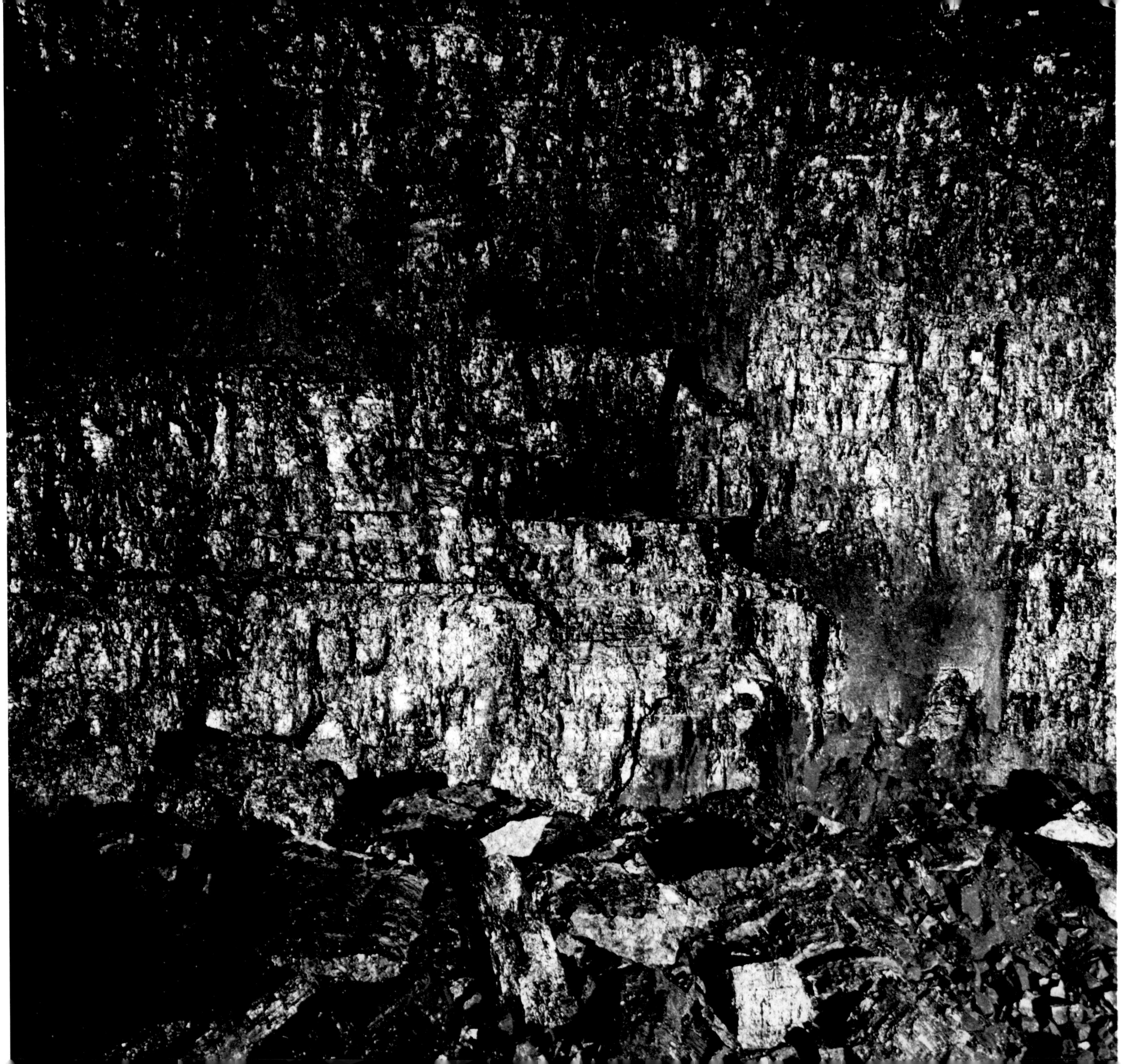

Face of a coal seam.

Overleaf: coal stockpile at coke ovens.

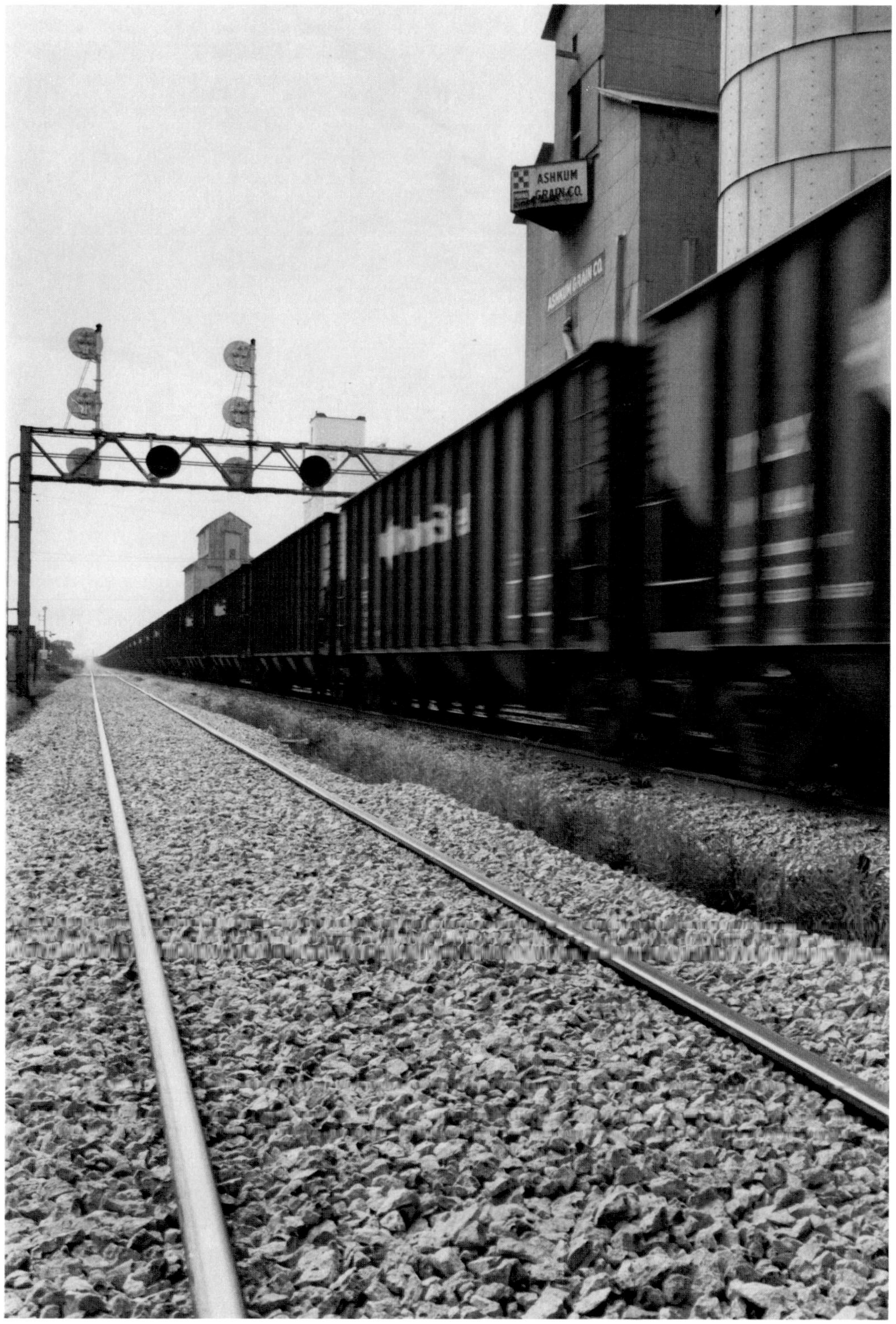

Unit coal train.

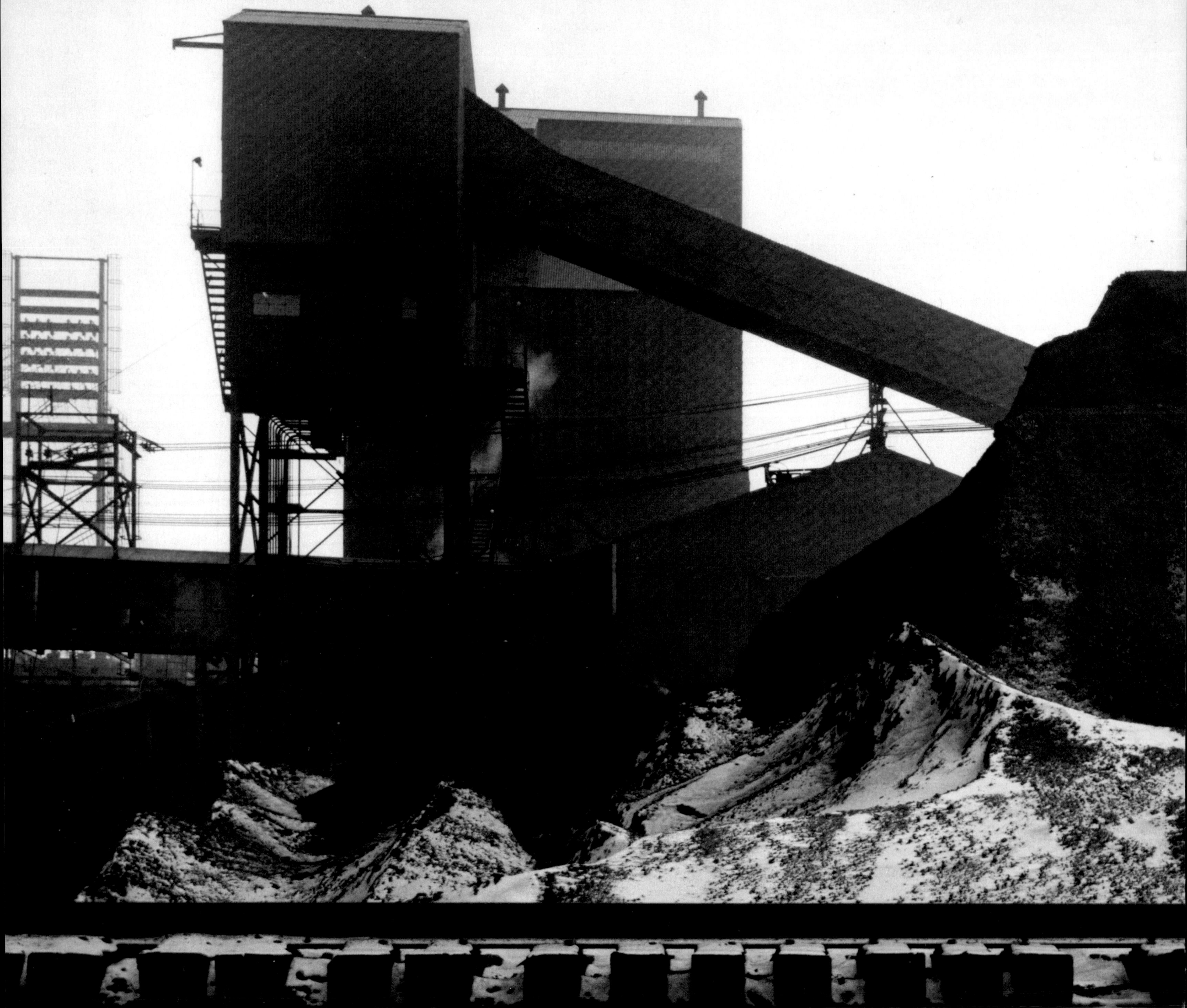

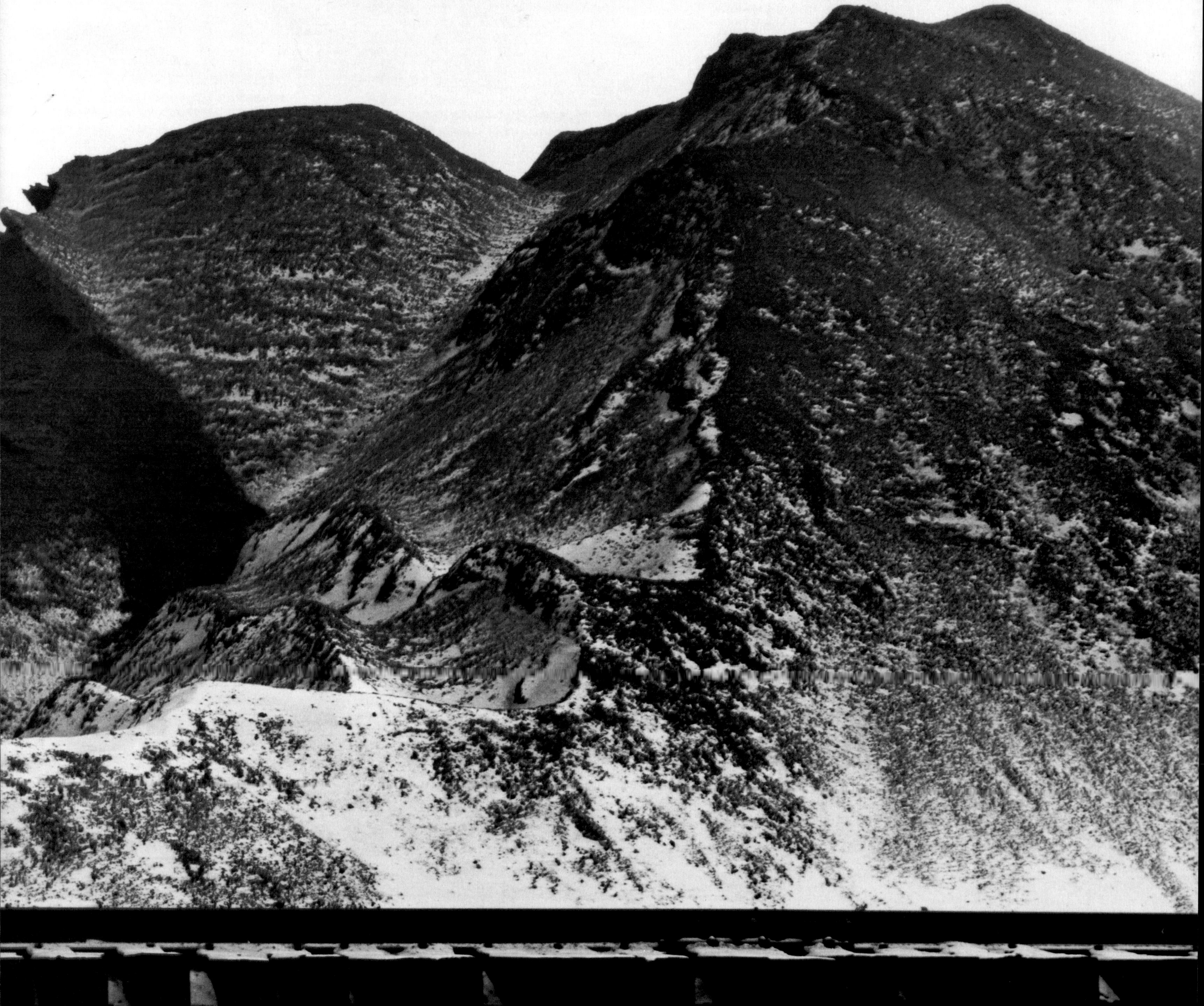

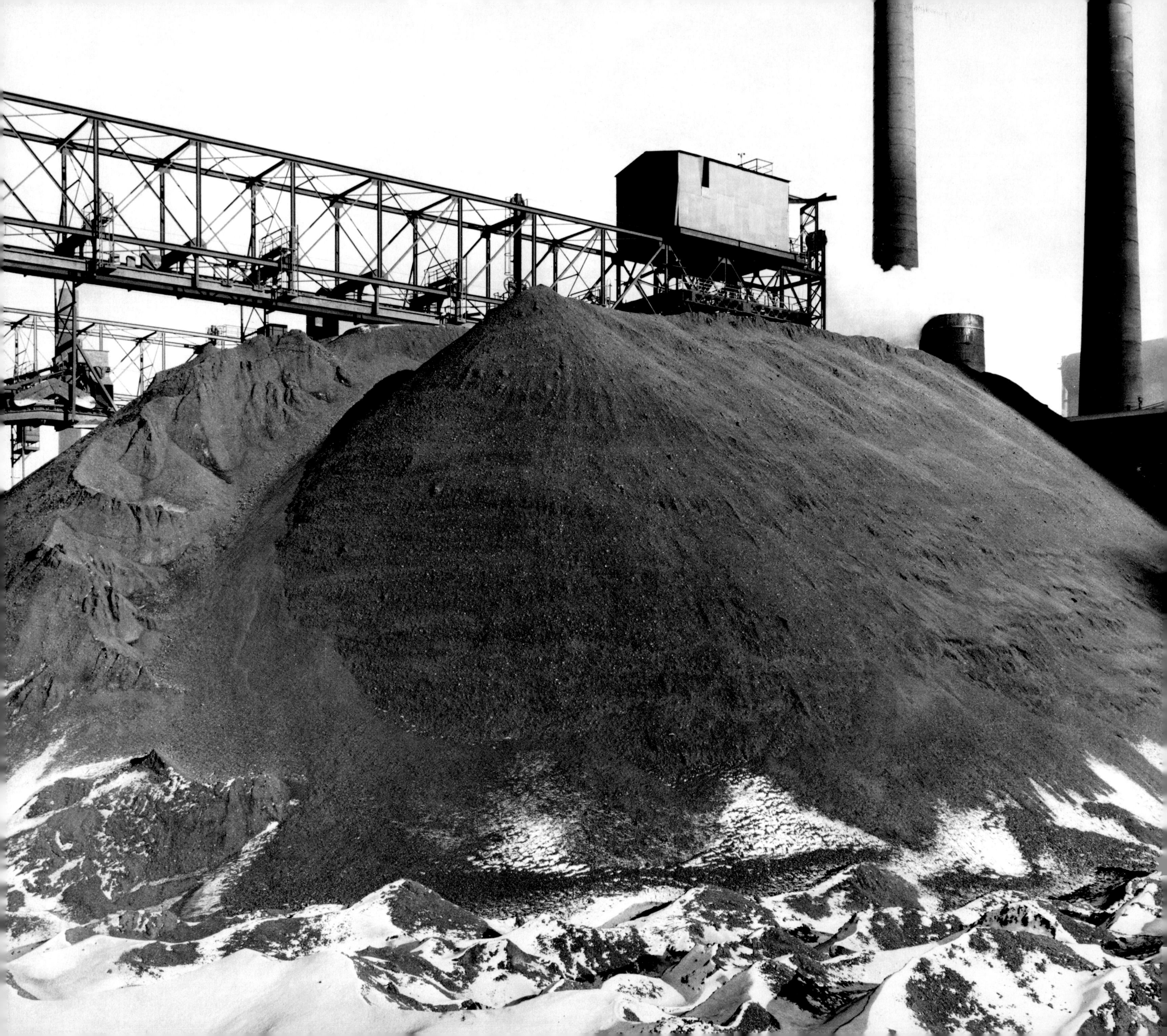

Coal stockpile
at coke ovens.

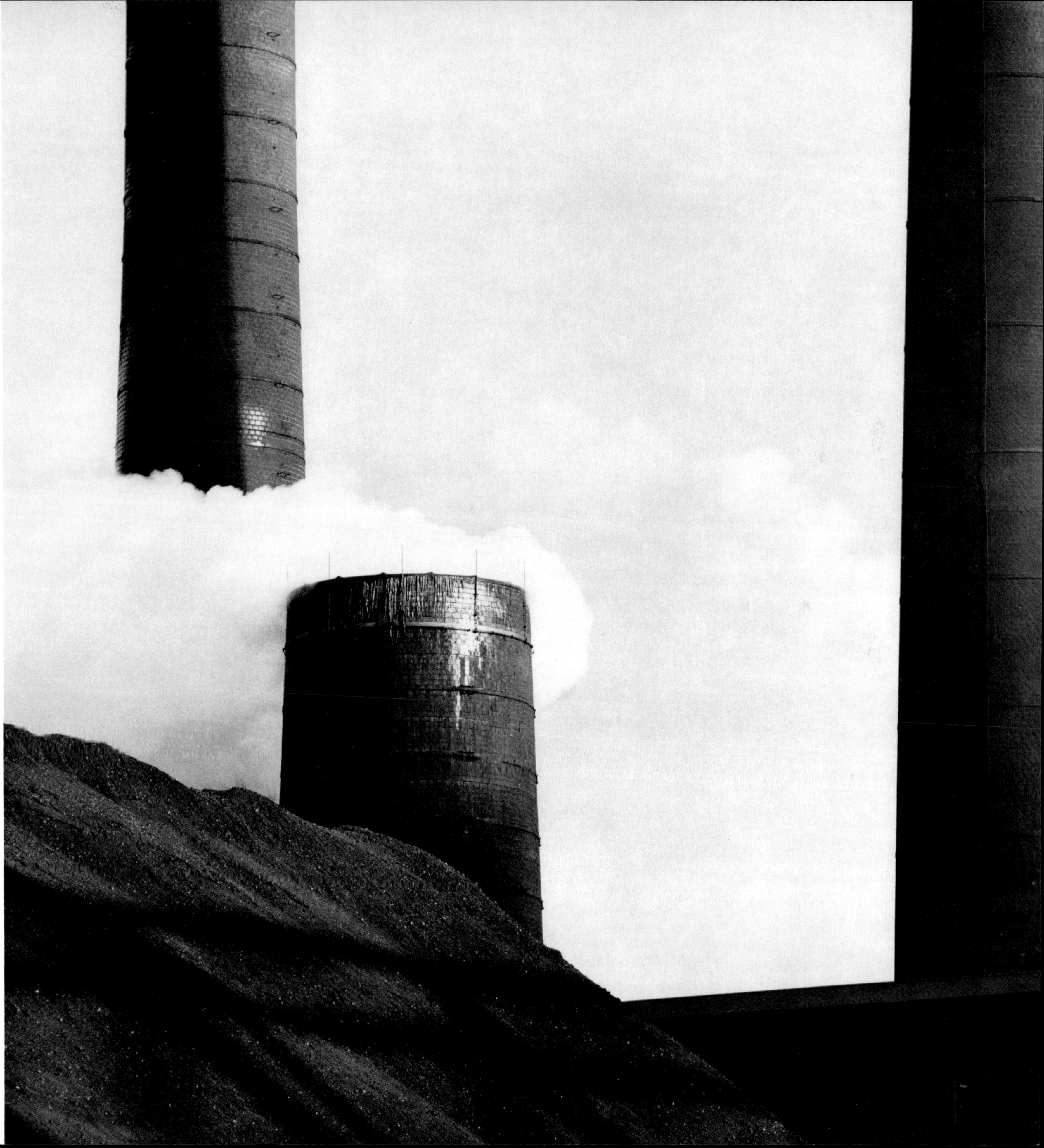

Detail of a
quenching tower.

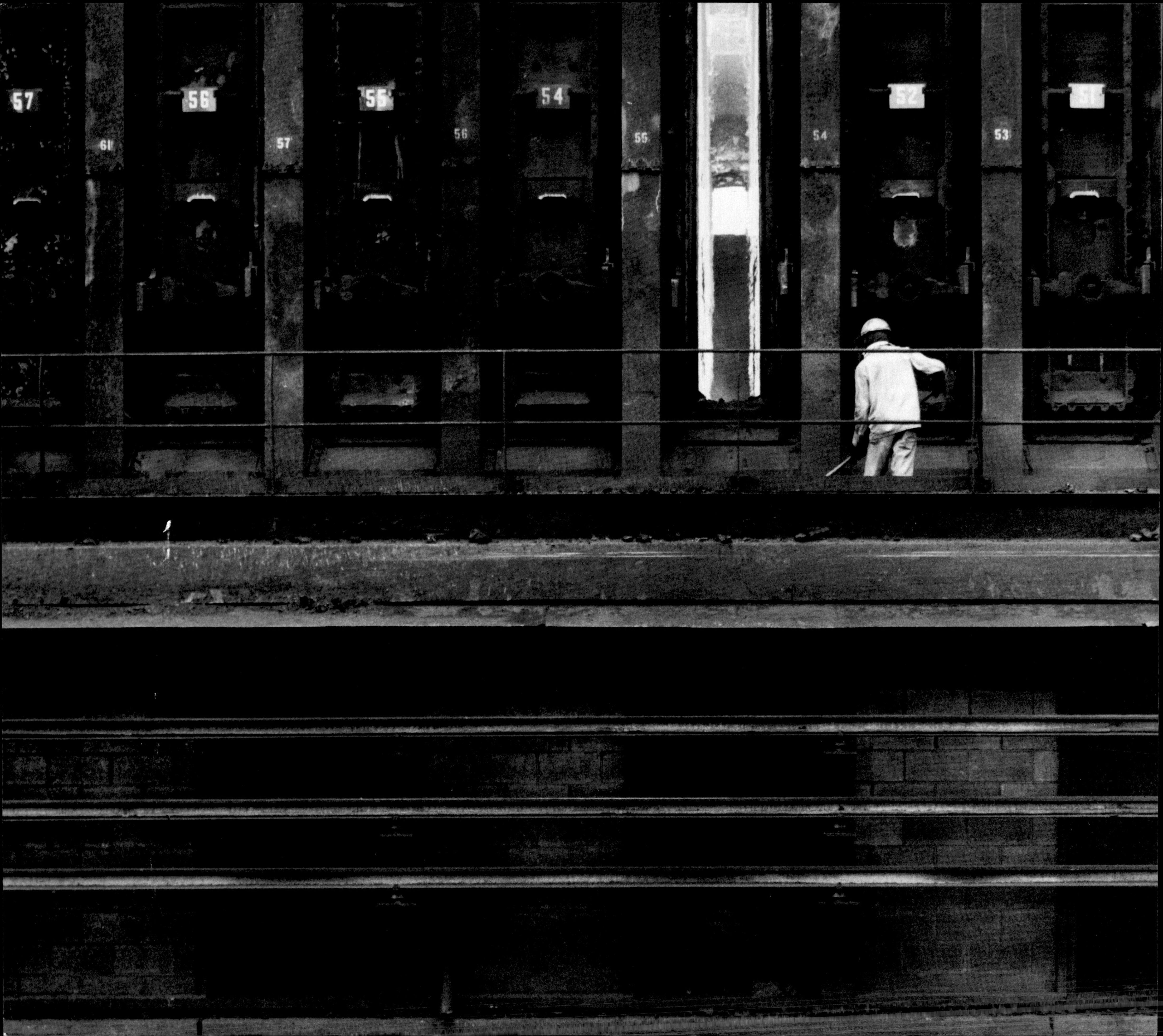
57
56
55
54
52
51
61
57
56
55
54
53

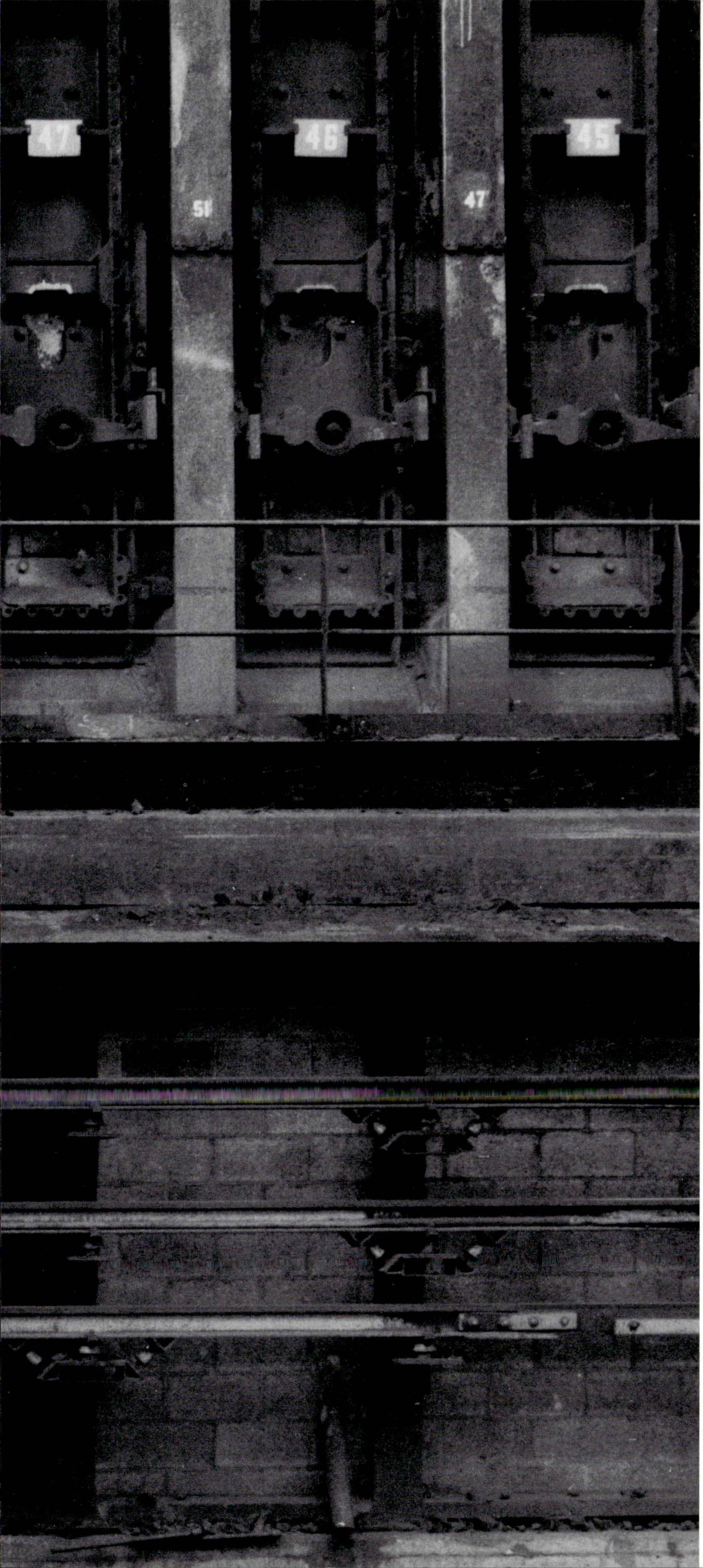

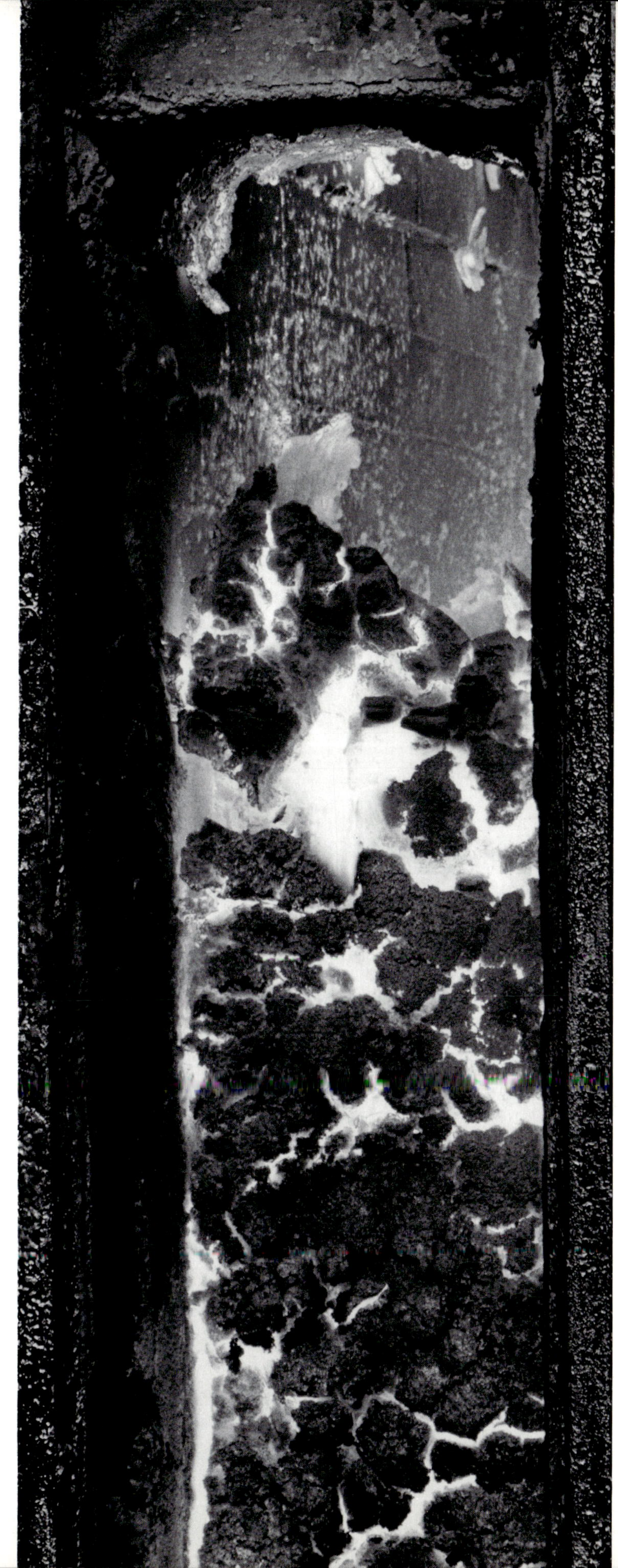

Coke-oven battery.

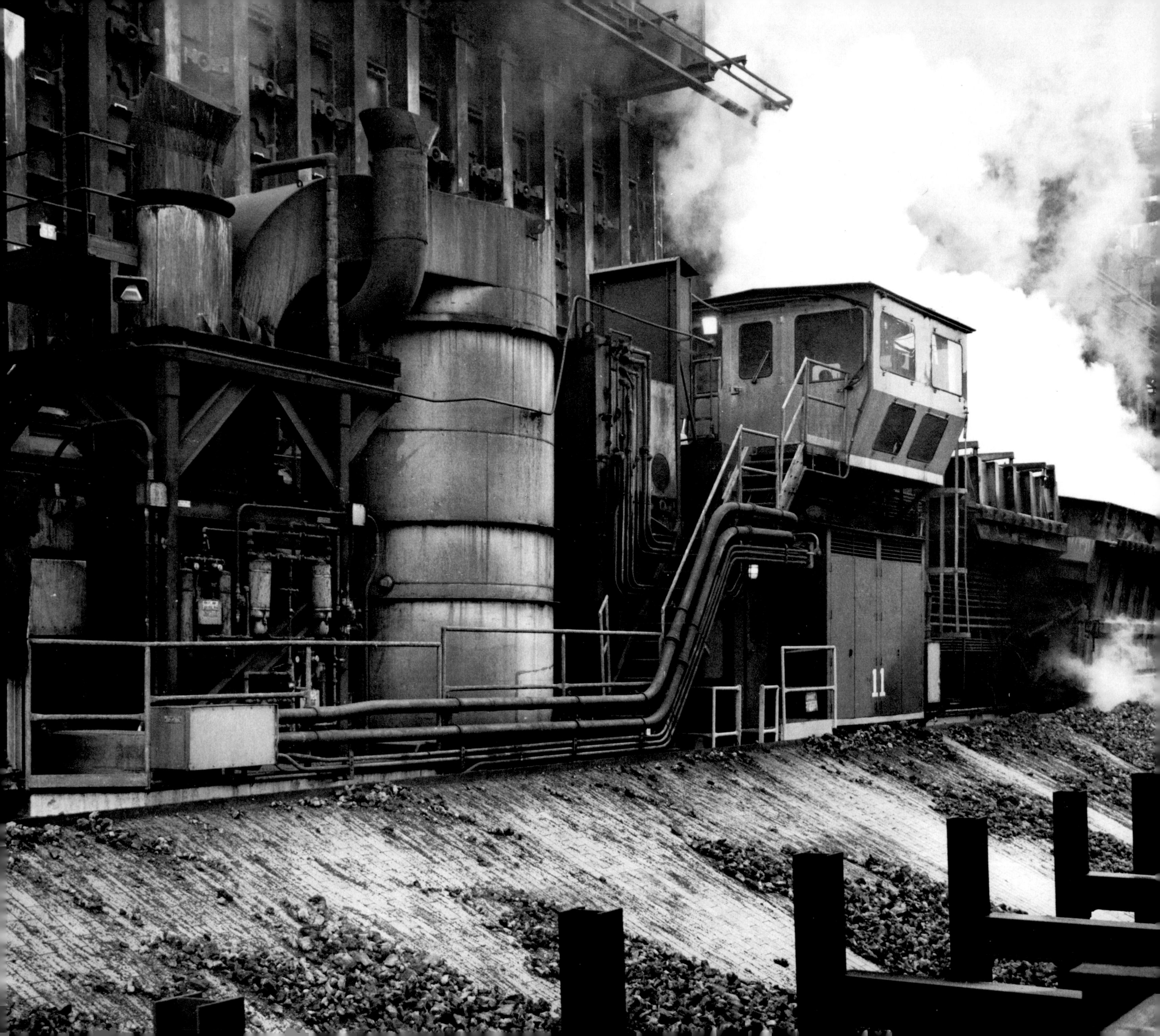
11

Quench car.

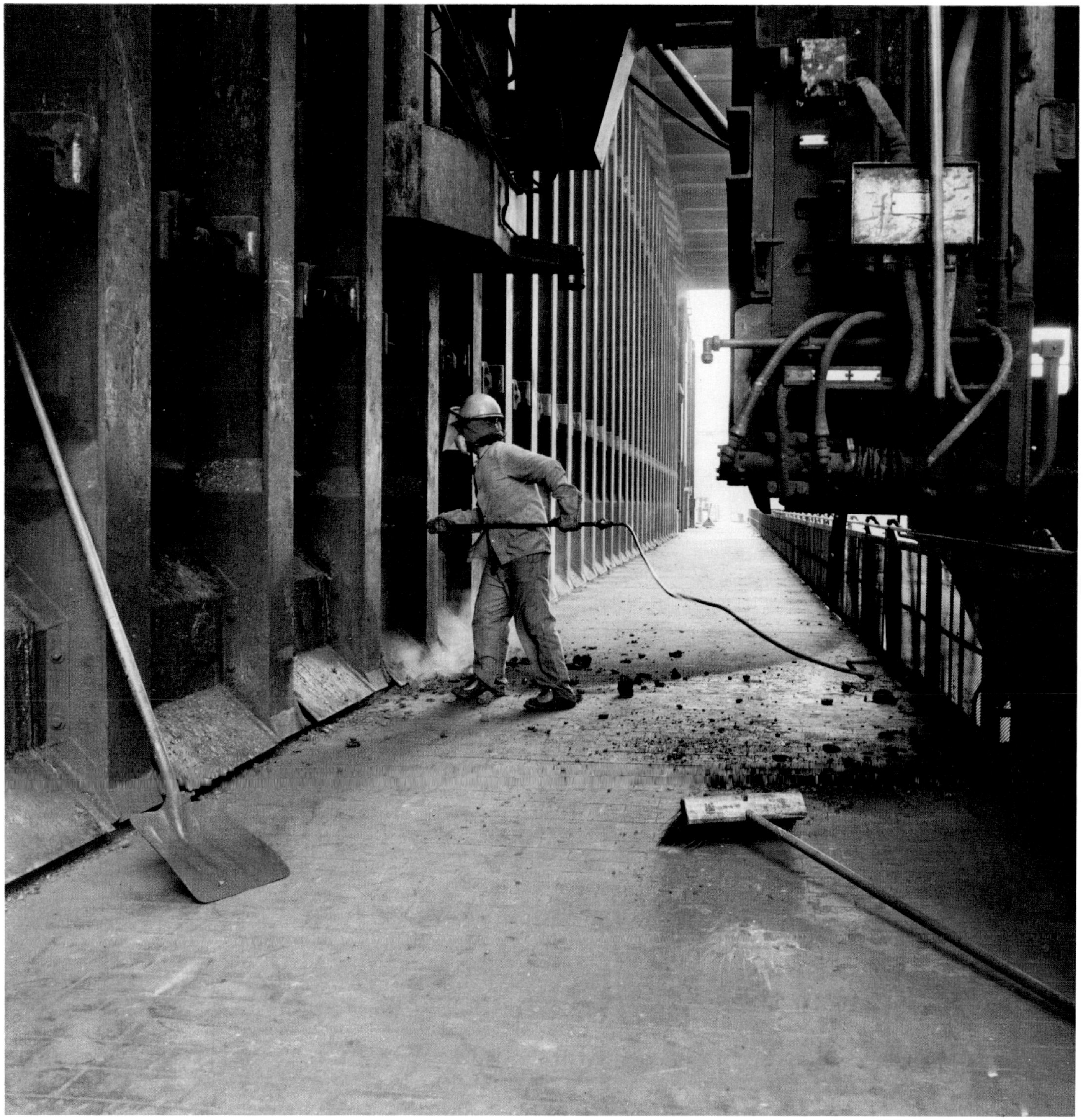

Coke-oven battery (from the "push" side).

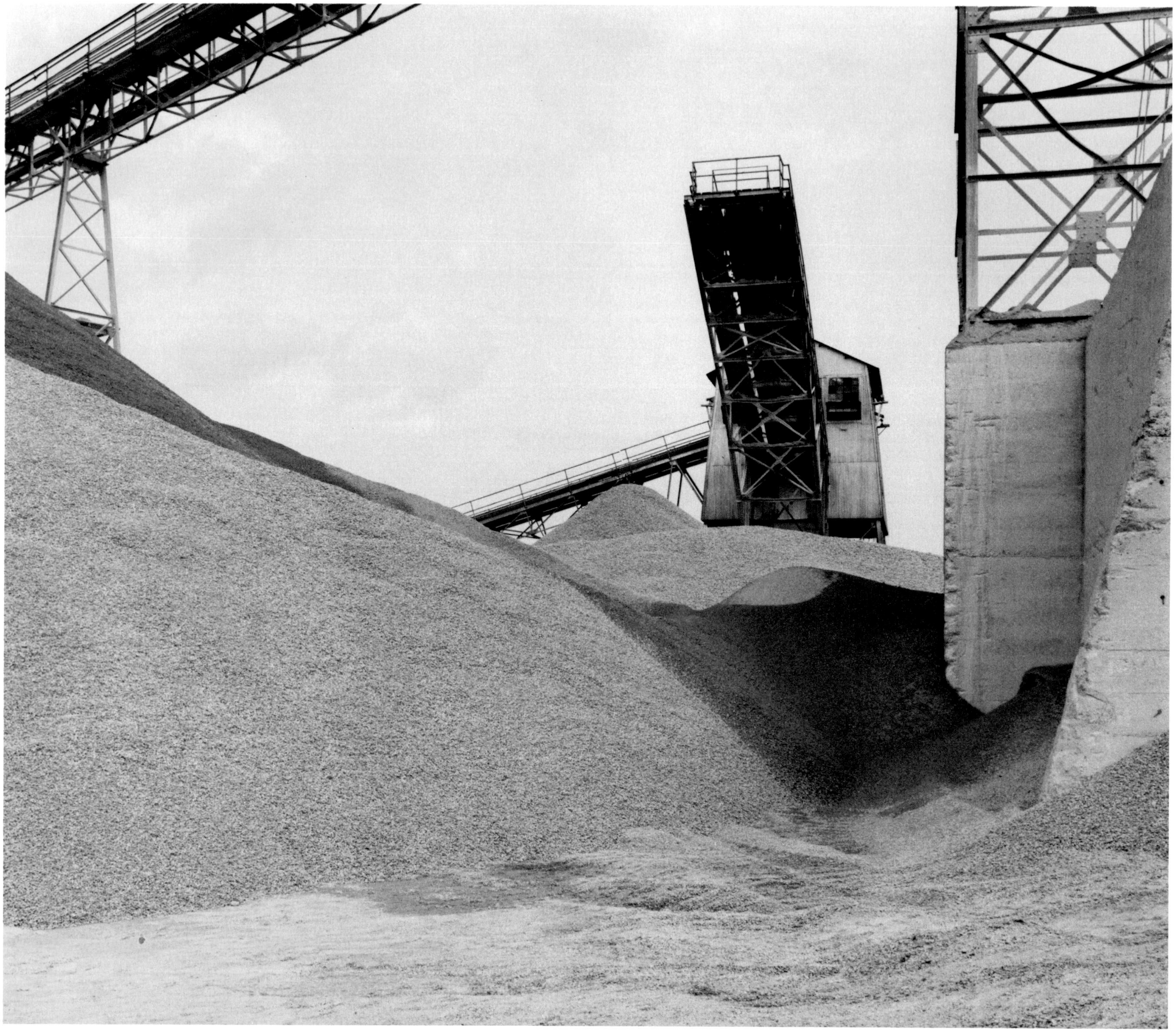

Crushers and sorters at a limestone plant.

Raw Materials: Limestone

Iron- and steel-making are basically refining processes, a series of purifying rites that free the metal from those unwanted "other elements" that are always found in combination with it. Paradoxically, the best way to remove them is by adding more rock to the mixture. High-calcium limestone and, to a lesser degree, dolomite are used as purifying agents. Once the limestone or dolomite melts in iron- and steel-making furnaces, it acts as a flux—it combines chemically with the unwanted impure elements to form slag. Slag, which is lighter than the metal, will float to the top of the molten bath, where it can then be drawn off.

Limestone is quarried in open pits similar to those used to mine taconite. The line between mining and quarrying is often finely drawn: quarrying, in the classic sense, refers to the extraction of stone per se from an open excavation; mining, whether above- or belowground, refers to the extraction of a more valuable material—ore, emeralds, and coal, for instance. Both open-pit mining and quarrying require essentially the same machinery.

Limestone and taconite are both blasted from ledges in the pit before they are processed further. Taconite, as we know, must go through a very complex and costly process before it can be fed into the blast furnace. On the other hand, the preparation of most limestone that is used by the steel industry is relatively simple: it is washed, screened, and sorted into piles of pieces of uniform size. The material can then be charged directly into the blast furnace for iron-making and into the open-hearth furnace for steel.

Recent changes in steel-making technology—the introduction of the basic oxygen process, for instance—require a faster-acting flux than limestone. Consequently, the industry is consuming increasing quantities of lime. Lime is simply limestone that has been fired, or "calcined," in kilns to drive off the carbon dioxide. Flux, whether lime, limestone, or dolomite, is a crucial ingredient in the processes of making iron and steel, and some thirty million tons of the material are required each year.

Most of the fluxing materials used by the American mills comes from the Great Lakes region, particularly Michigan and Ohio. The bulk of it travels over the same waterways as the iron ore and in the same type of vessel, the laker.

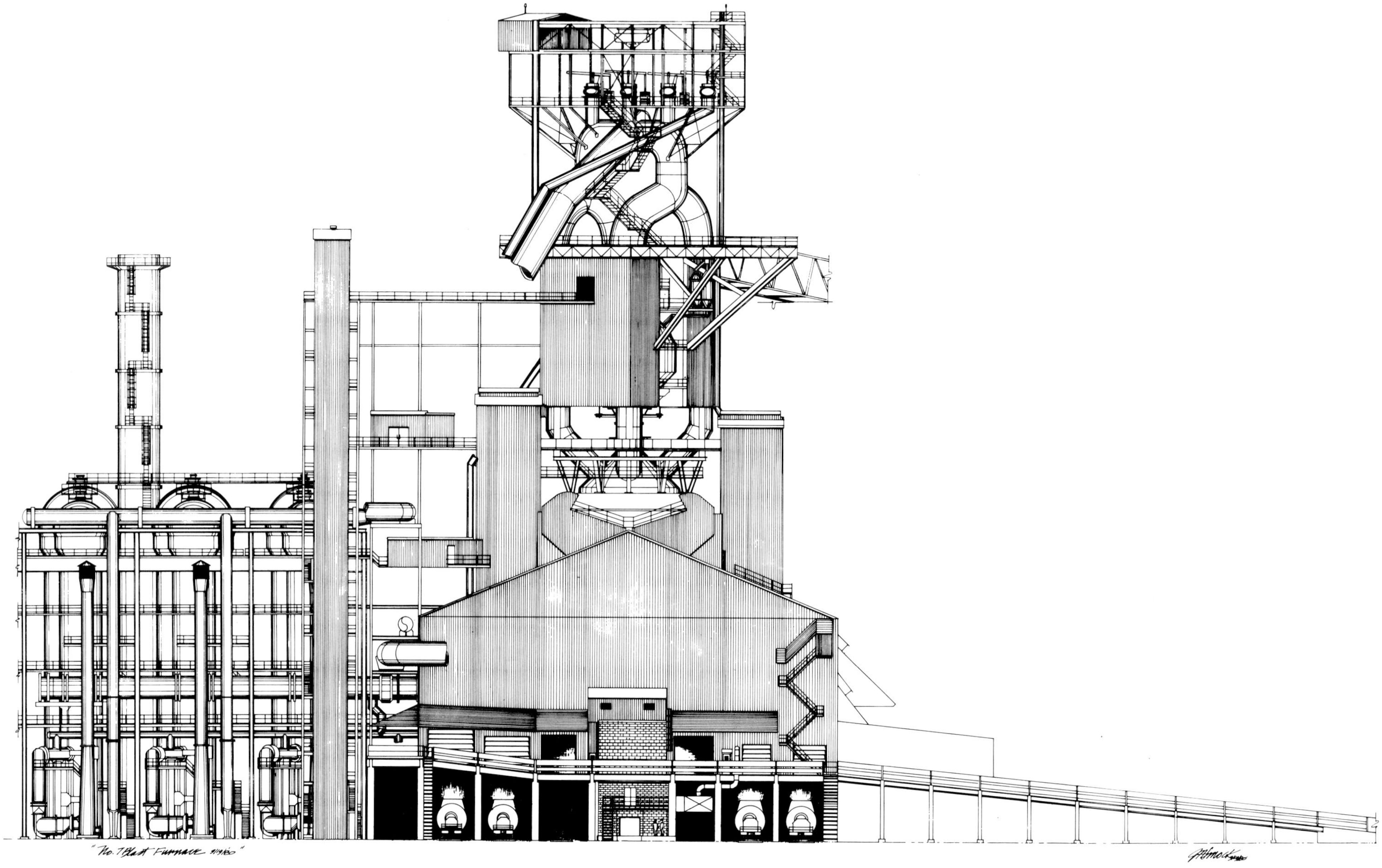

Blast furnace, cross section of new model.

The Making of Iron

Up to this point in the narrative, and in the steel-making process itself, the focus has been on the gathering and processing of raw materials. Now the curtain goes up on Blast Furnace Row. Often rising as high as a twenty-story building, behind mountains of taconite pellets and limestone, are the blast furnaces, the steel industry's most fantastic architecture. Blast furnaces, with their satellite structures—stoves and antipollution devices—and all of the connective piping, present a scene without counterpart in the industrial world.

The furnace's awesome visage, however, belies the basic simplicity of its operation. The art of iron-making may have evolved into a very precise science and blast furnaces may have become ever larger and ever more efficient, but man has been using the same principles implemented in the blast furnace to make iron for six hundred years. Within each furnace, alchemy is at work: primal matter—ore, stone, and coke—coalesce and melt. The result is white-hot liquid: iron.

The blast furnace is the steel industry's smelter, the place where the metallic iron is freed from the ore. The structure, in simple terms, is a vertical steel cylinder lined with heat-resistant refractory brick in which raw materials are melted down by a blast of hot gas. Iron-ore pellets, coke, and limestone are measured in a stockhouse adjacent to the blast furnace before skip cars carry them to the top of the furnace and dump them in. The standard recipe, or charge, calls for five hundred pounds of limestone (or a similar flux), one thousand pounds of coke, and thirty-three hundred pounds of iron-bearing material. (In some of the larger blast furnaces of today, the volume of the charge is much greater, but the ratio of raw material remains essentially the same.)

The huge amount of hot air needed for the blast—as much as four and a half tons in the manufacture of one ton of molten iron—is heated by exhaust gas from stoves located adjacent to the furnace. Each furnace requires a minimum of two of these tall, cylindrical stoves: while one is "on gas" and storing heat, the other is "on blast" and supplying air to the furnace.

The pressure within the furnace is enormous, and the charge is prevented from being blown back out of the furnace by a system of valves, or "bells." The material passes over the first bell, which is then closed. The second bell beneath it is lowered, allowing the charge to drop into the furnace; after this step is completed, the second bell is raised into place. Gravity then takes over, carrying the mass slowly downward. During their descent, the materials are preheated by hot gases rising within the furnace. As the mass nears the bottom it is met by a continuous blast of superheated gas blown upward from the base of the furnace through openings called tuyeres (pronounced "tweers"). This blast, from which the structure takes its name, causes the coke to burn with great intensity. The extreme temperature—at this point above 3,000 degrees F.—brings about "smelting," the release of pure iron from the ore. In order to further increase the heat, coal dust, natural gas, or fuel oil is often injected through the tuyeres. About halfway down the stack, the limestone begins to melt and react with the impurities in the ore and coke to form slag. At the same time, some of the silica in the ore and the carbon in the coke are reduced and dissolved in the iron. At the bottom

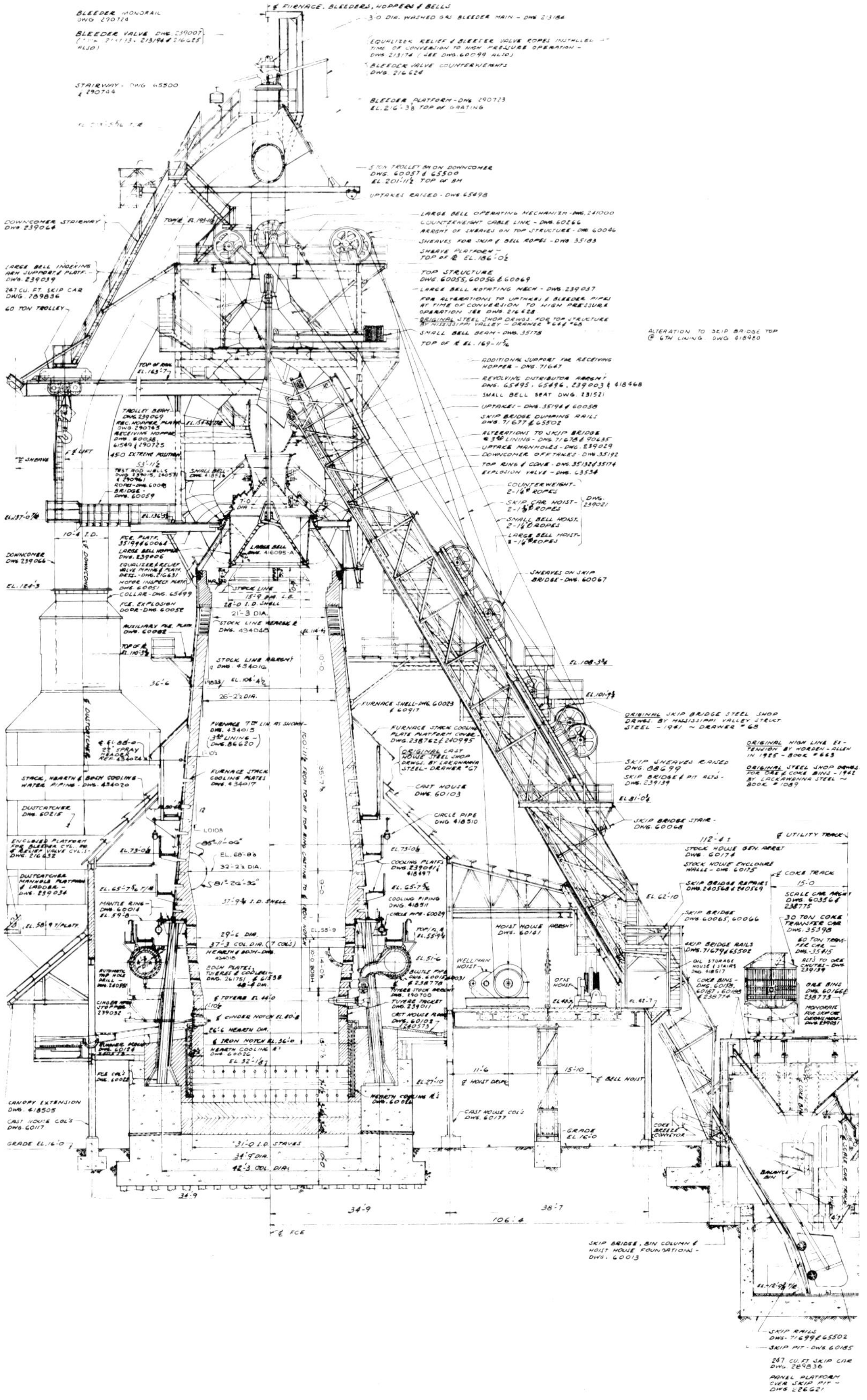

Blast furnace, cross section.

of the stack, called the "hearth," a pool of white-hot molten iron begins to collect. On top of this floats the slag. When the pool reaches a depth of four or five feet (which, coincidentally, happens about every four to five hours), the furnace is tapped, or "cast," by drilling out the plug in the top hole at the hearth. At this point the slag is drawn off through a "slag notch" above the level of the iron.

Here, at the base of the great furnaces situated in the cast houses, man is confronted with molten metal for the first time. A beautiful pale orange stream of liquid iron courses without a sound (it contains no oxygen to make it bubble) through channels in the sand, downward from the hearth to the cars that wait below to carry it away to the steel mill.

Iron made in the blast furnace is known almost universally as pig iron, a quaint sobriquet derived from the early days of iron-making. At one time the molten metal ran from the hearth through a central channel, which led into a series of smaller channels on either side of the larger one. From these smaller channels the metal flowed into holes in the sand, where it was allowed to cool. To a largely agrarian society, the arrangement was reminiscent of a sow suckling her newborn. The nomenclature has endured—today the main channel from the tap hole is a "sow," and the metal that pours through is still often referred to as "pig" iron.

Most pig iron—or hot metal, as some of a more pragmatic nature prefer to call it—is destined for the steel-making furnaces. In the United States this amounts to some eighty million tons annually. Less than five million tons—a small amount by comparison—is sold as "merchant pig iron" to foundries, where it is cast into such diverse objects as automobile-engine blocks, water and sewer pipes, and dutch ovens.

The pig iron bound for a rendezvous with a steel furnace is poured into specially designed railroad cars known as bottle cars, or pugh cars. These cars, like everything else that molten metal might come into contact with, are lined with refractory brick, which in this case acts as an insulator. Slag, on the other hand, is poured into another type of car, unceremoniously called a slag pot. Slag is essentially waste, and most of it ends up in the dumps. A small amount, however, is used as railway and highway ballast and, occasionally, in the manufacture of concrete.

Iron-making takes place around the clock, seven days a week, year in and year out. Blast furnaces, once fired up or "blown in," operate continuously, often for a period of several years.

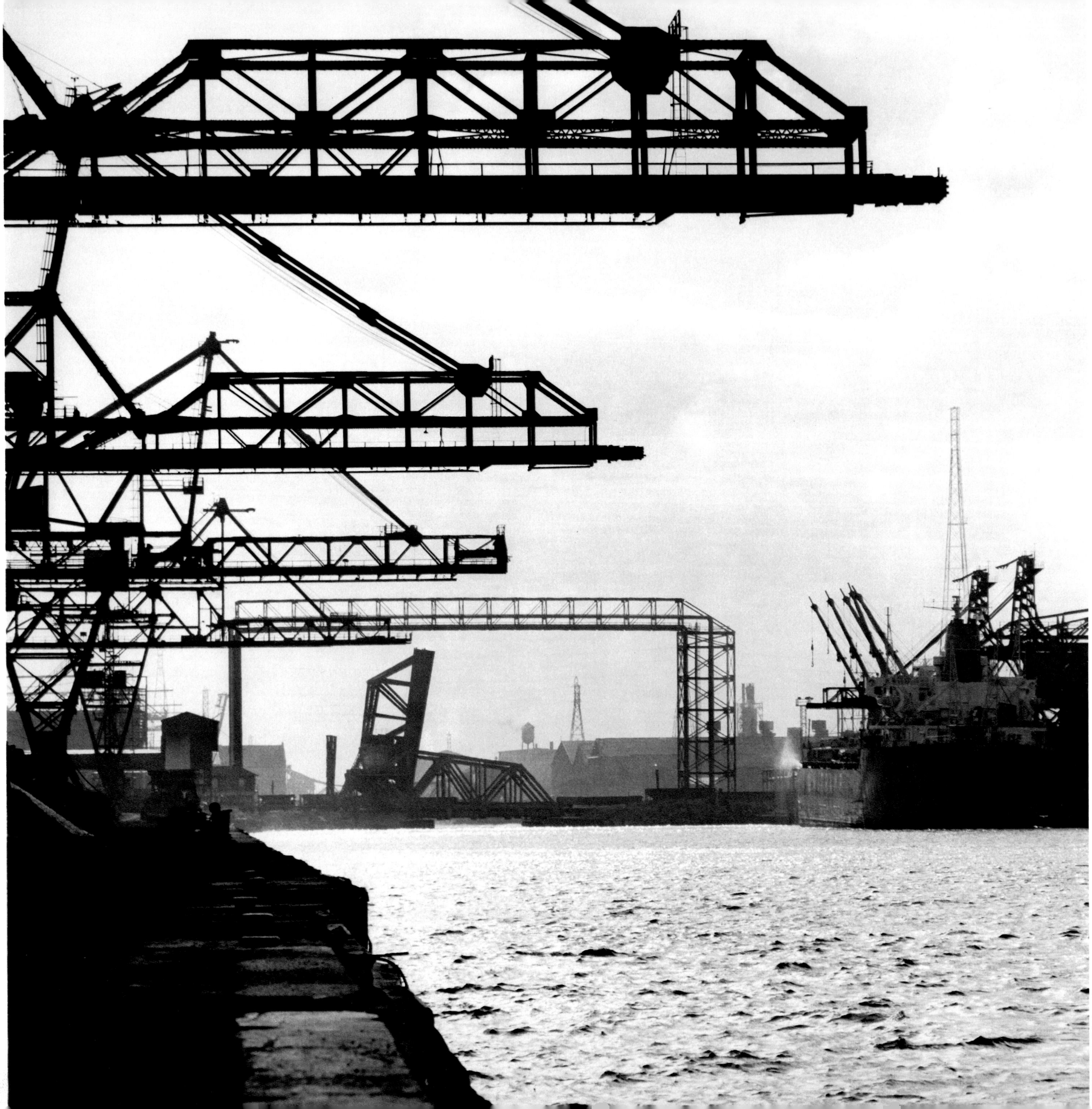

Steel mills.

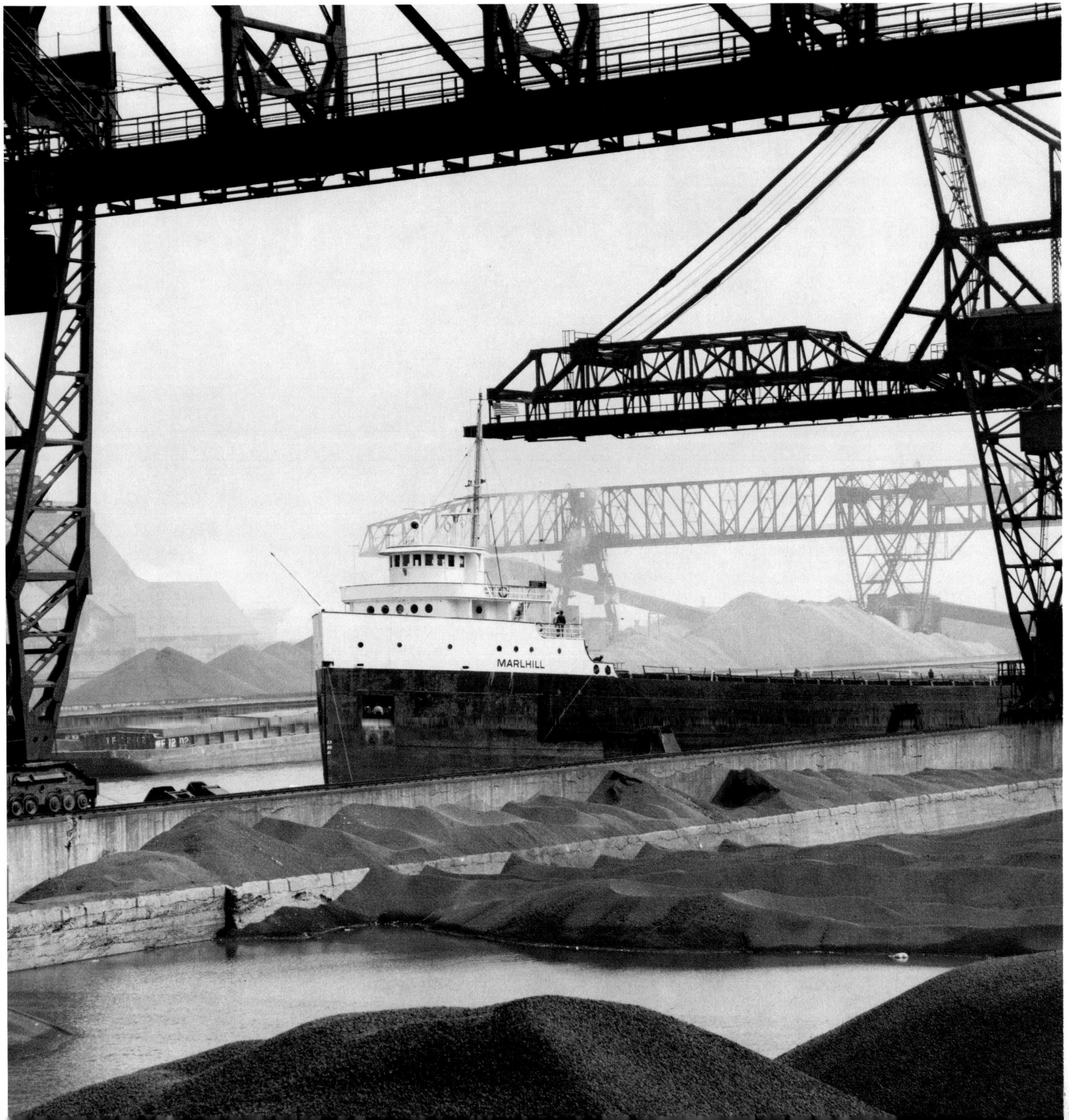

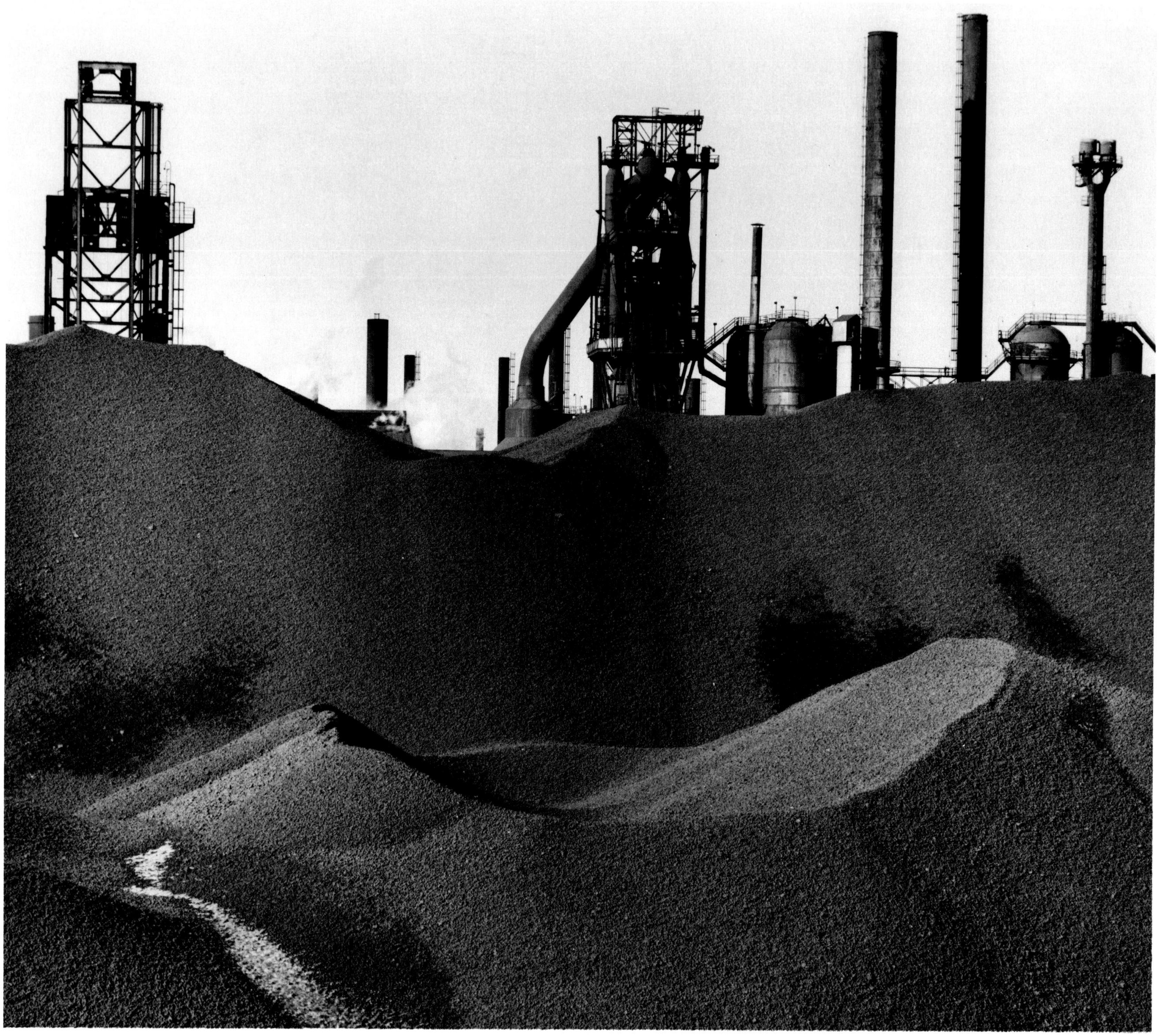

Blast furnaces and stockyards.

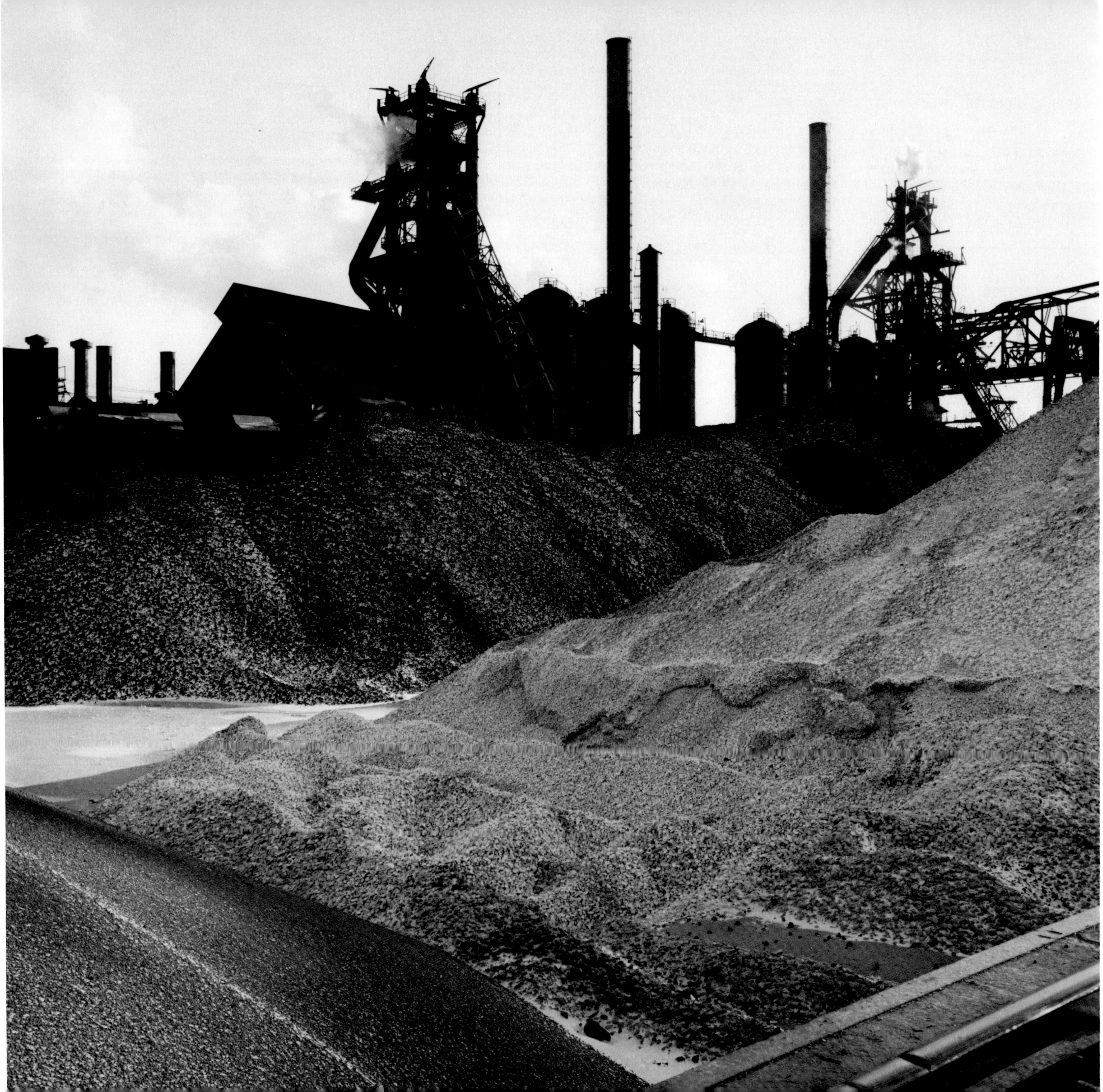

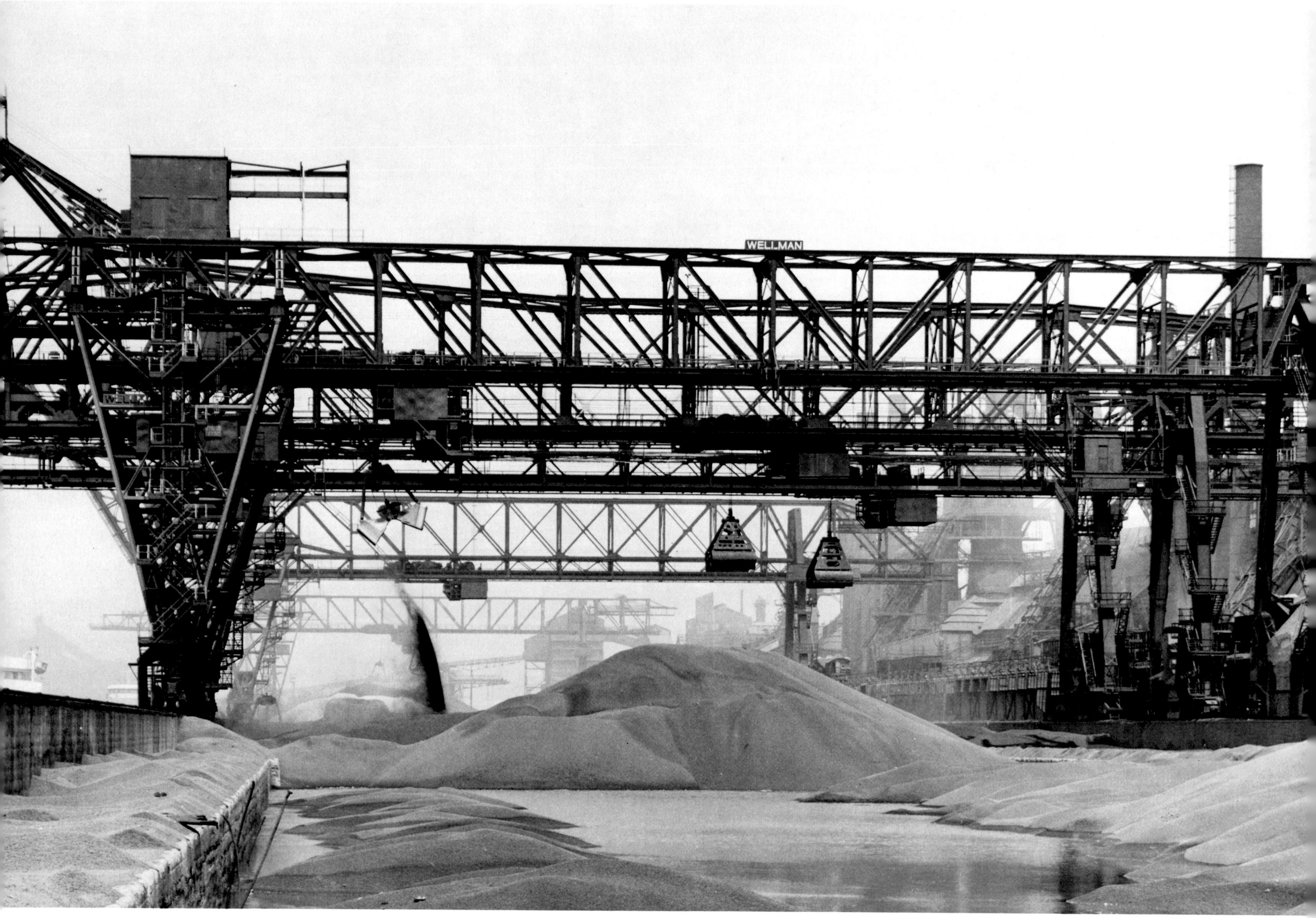
WELLMAN

Ore bridges.

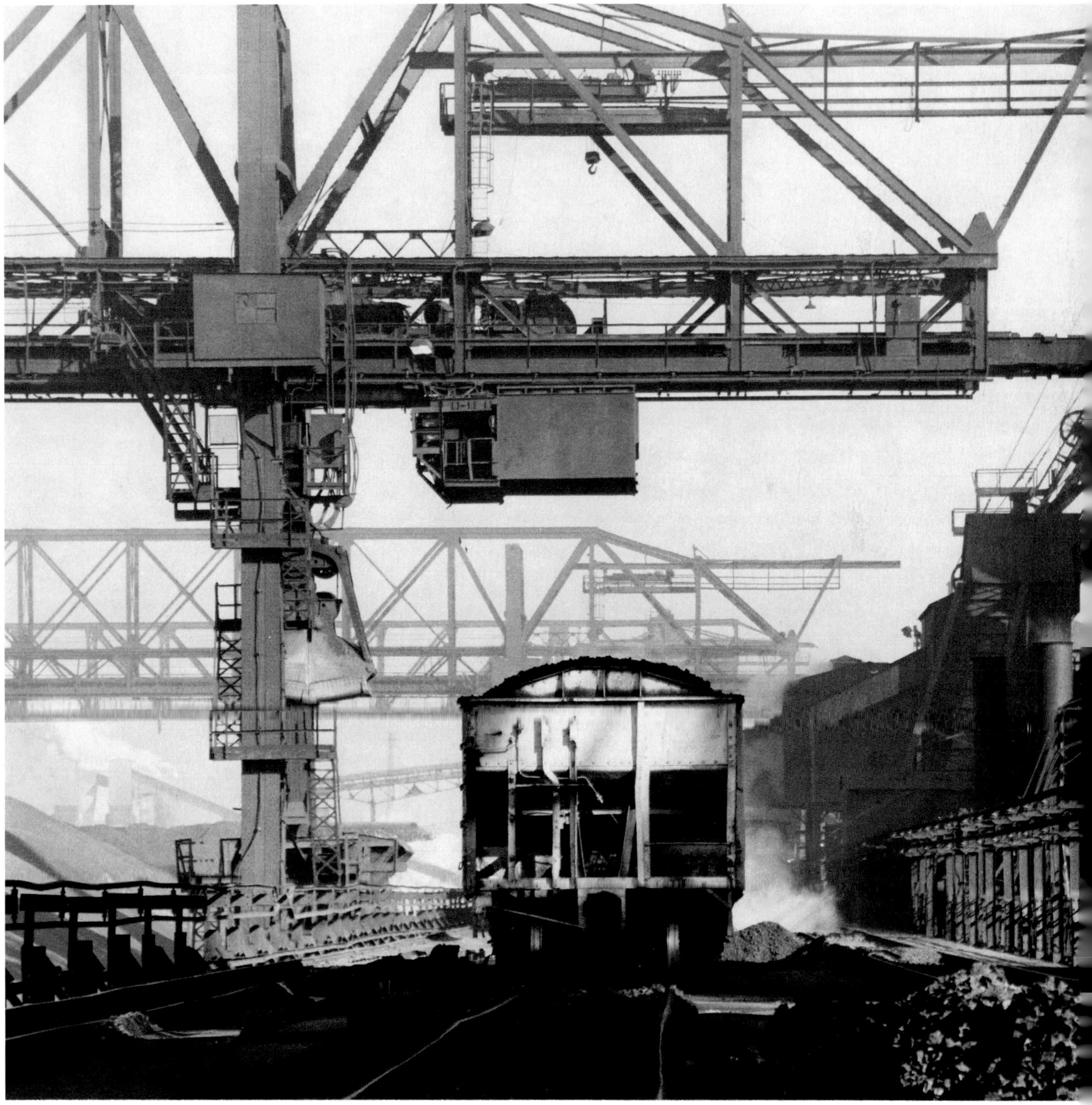

Blast furnaces.

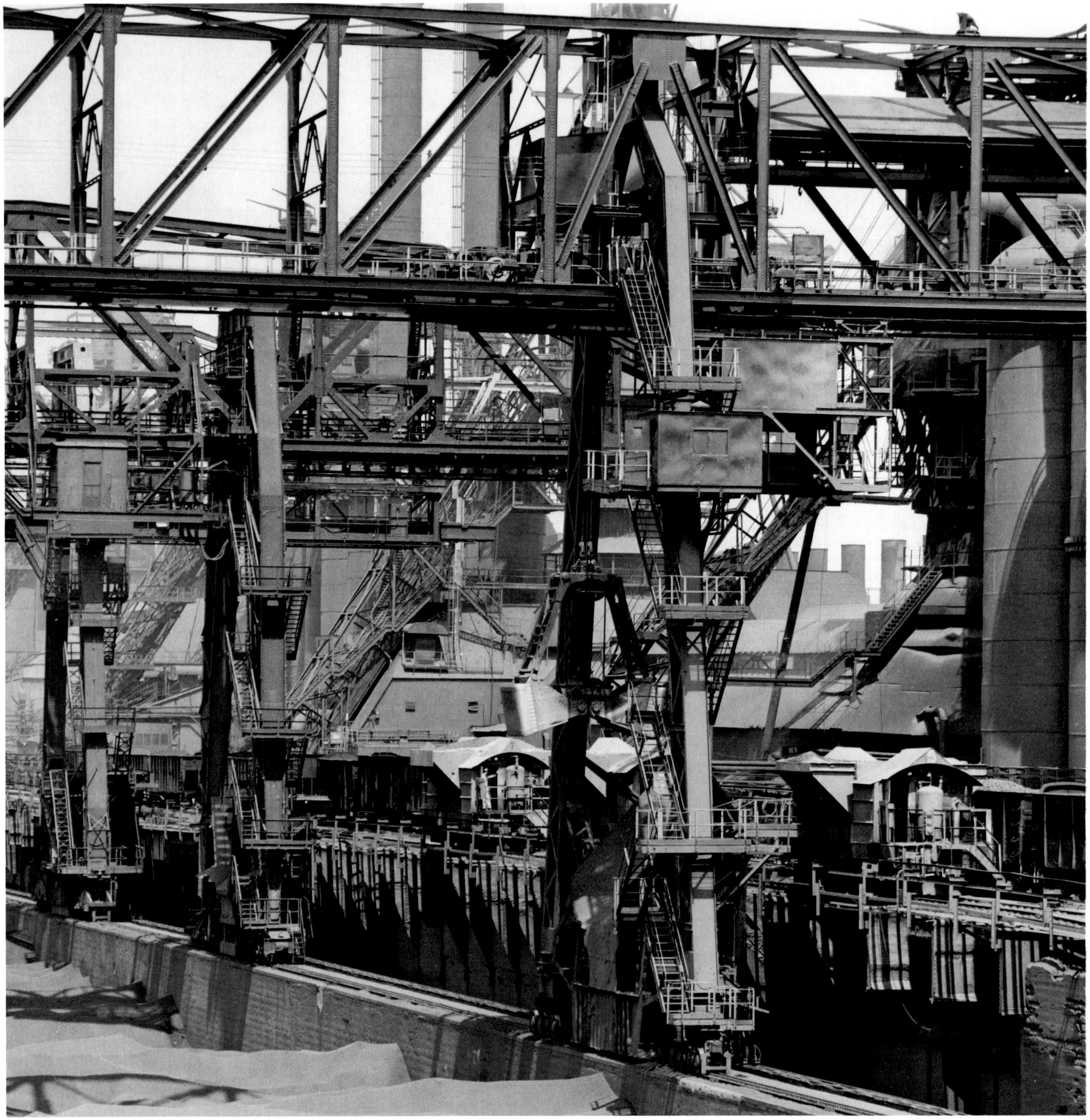

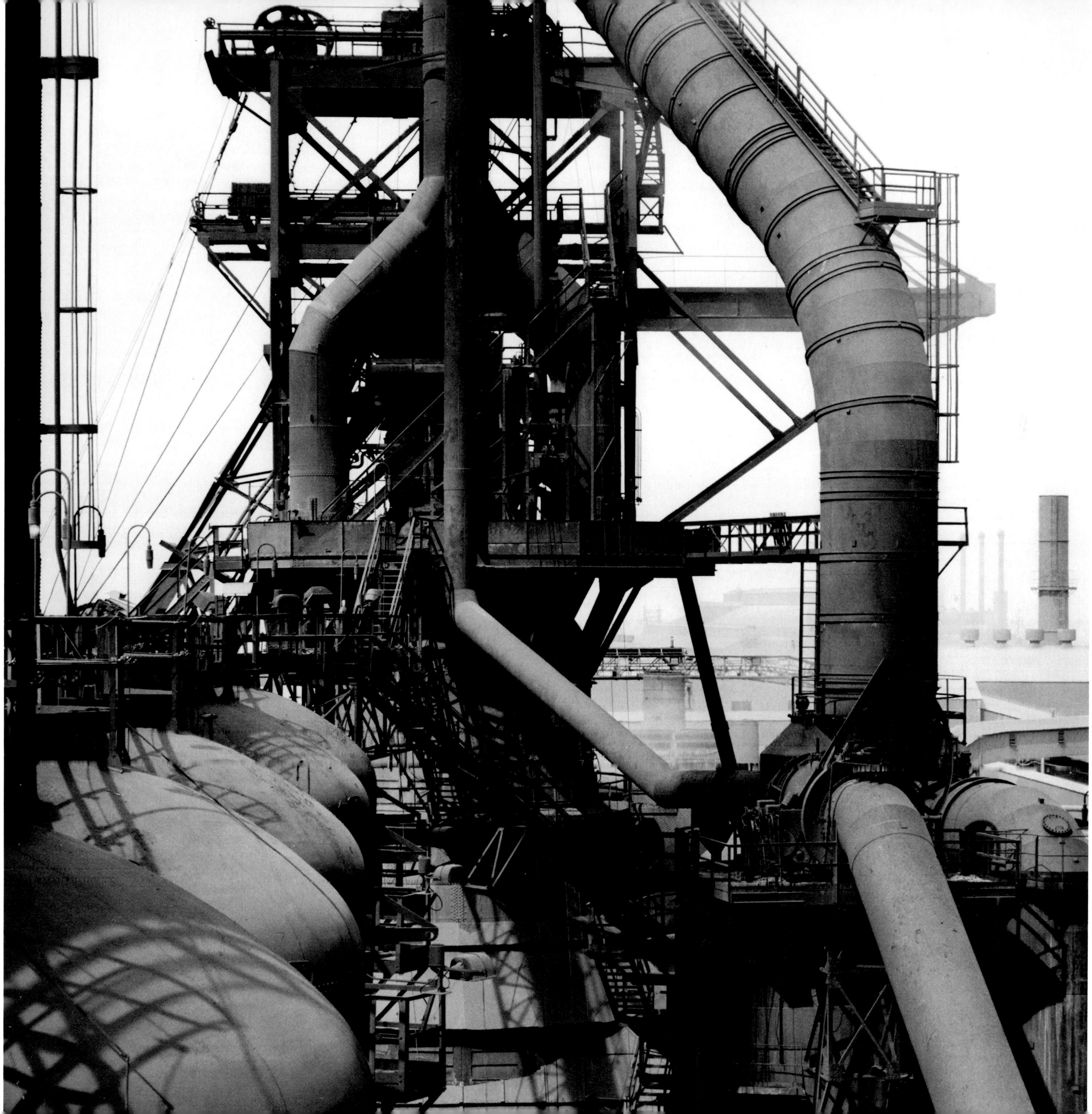

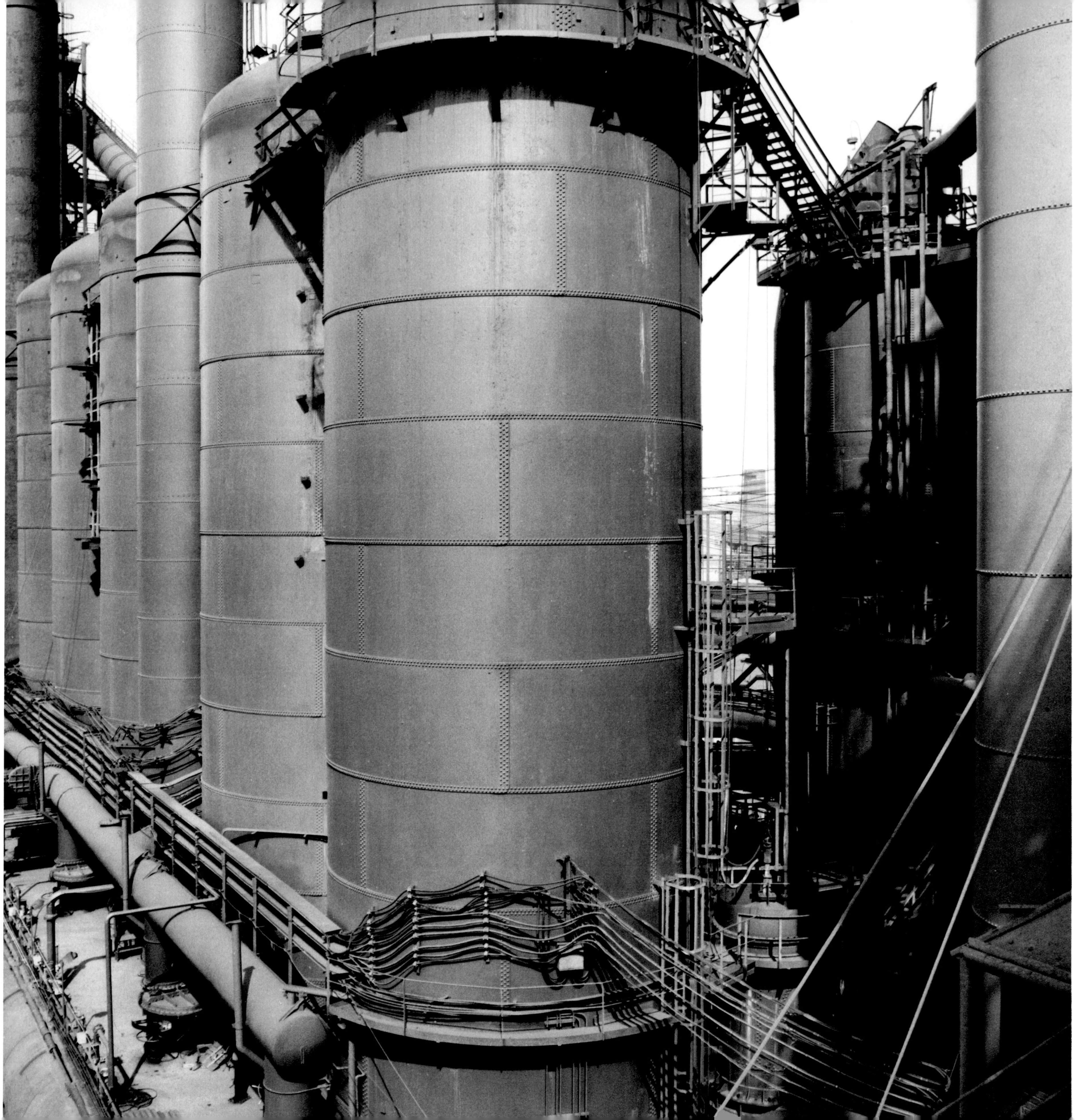

Blast-furnace stoves.

Cast-house operation.

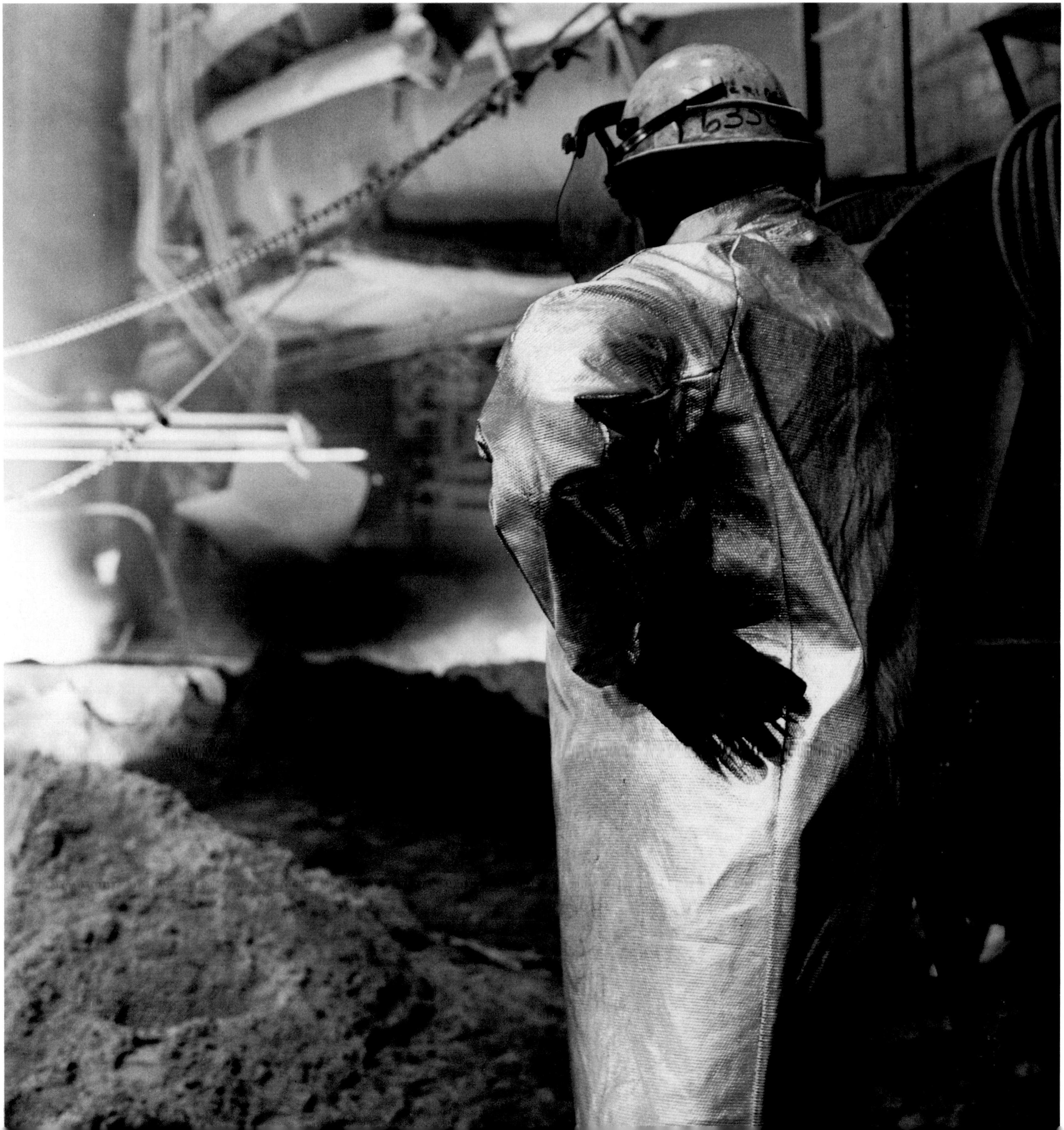

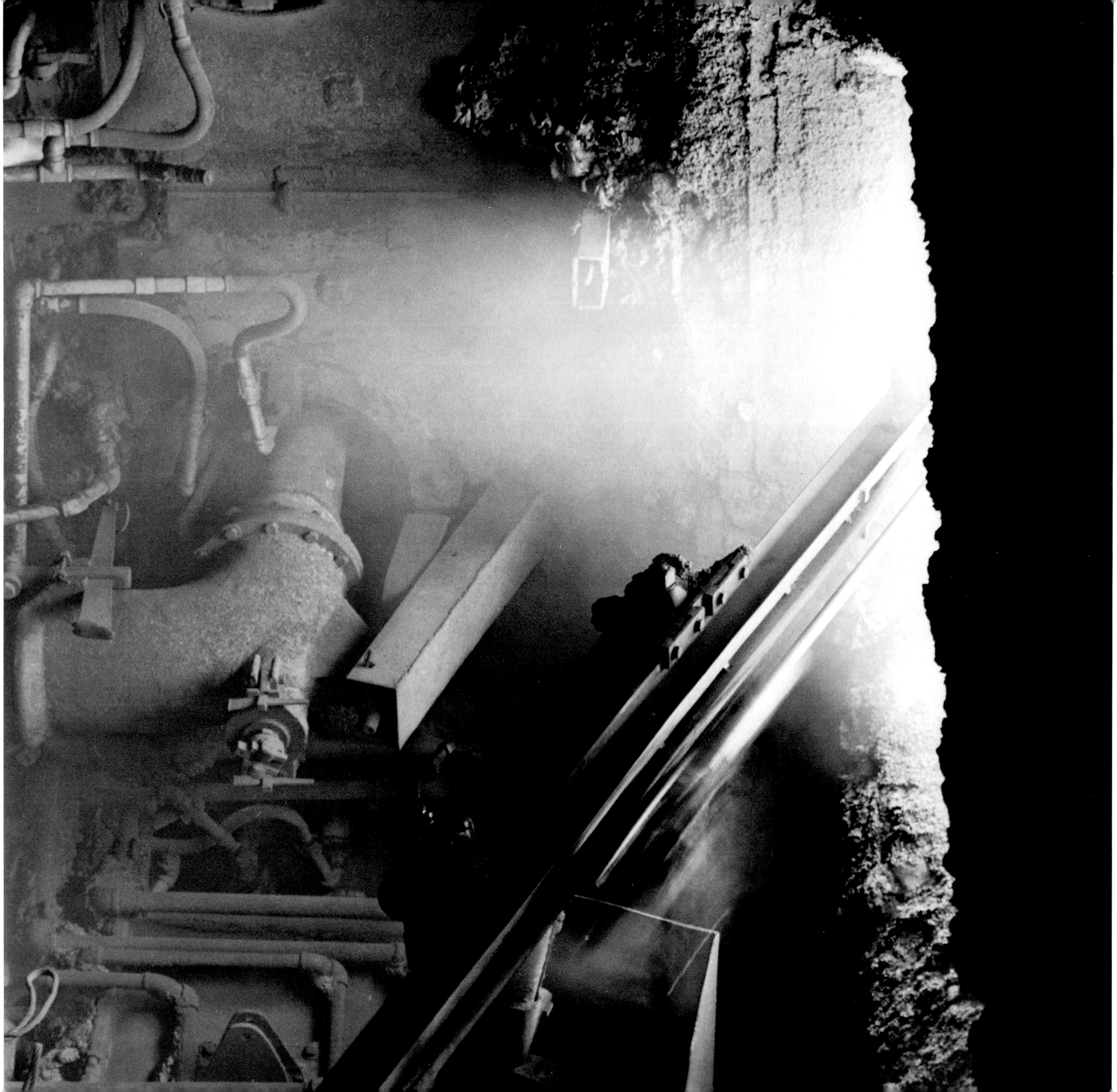

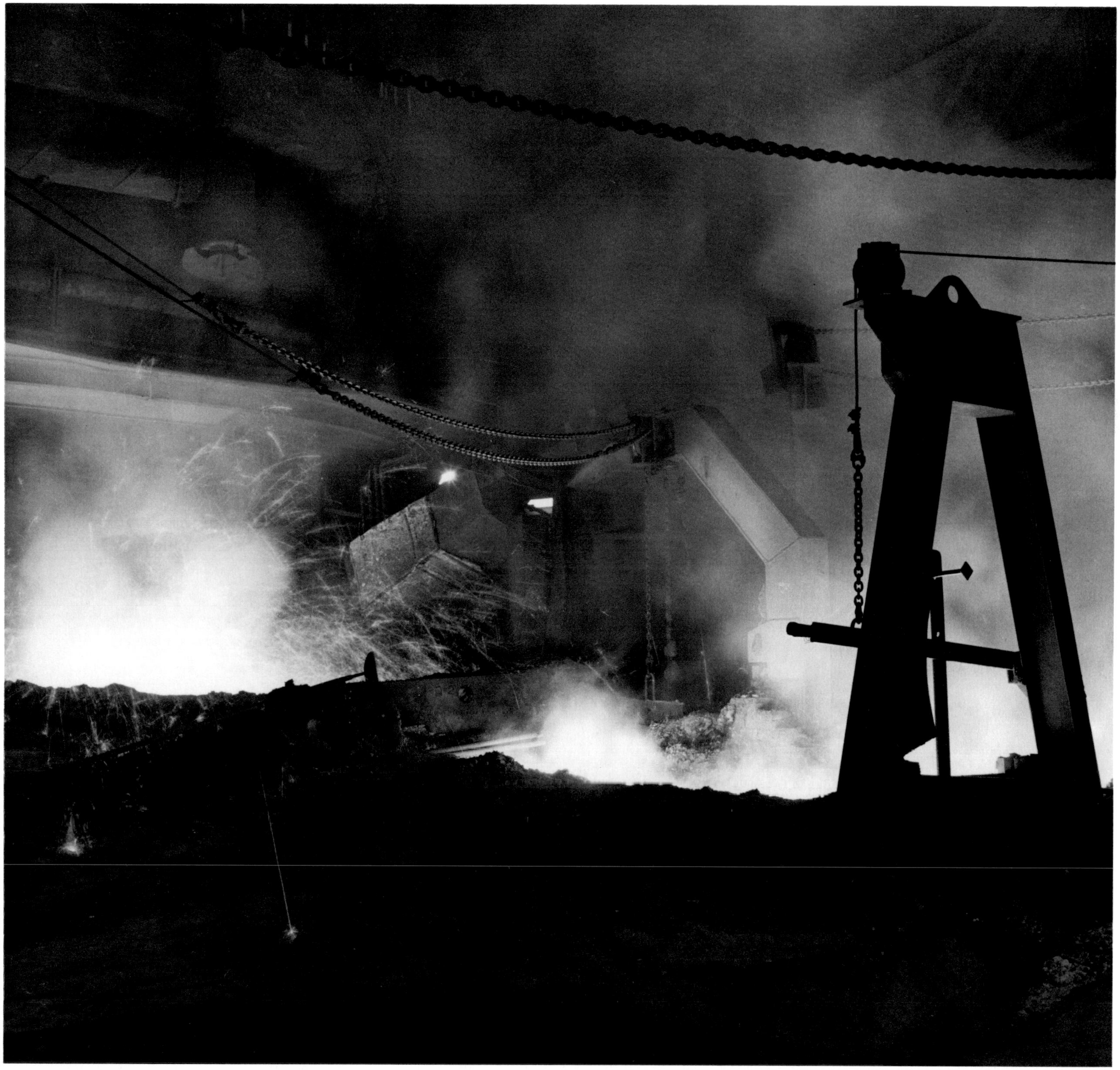

Casting in a blast furnace.

Casting.

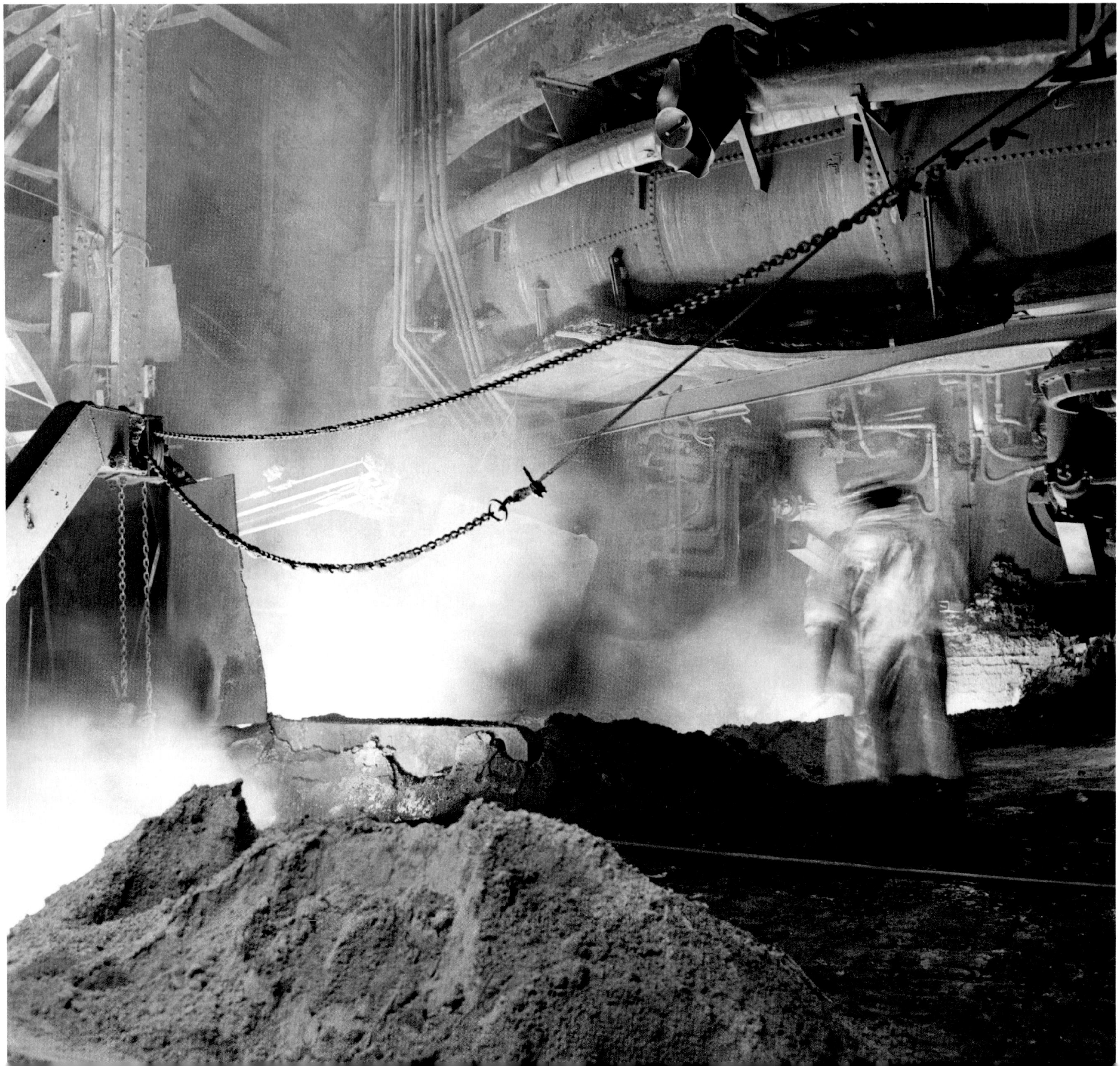

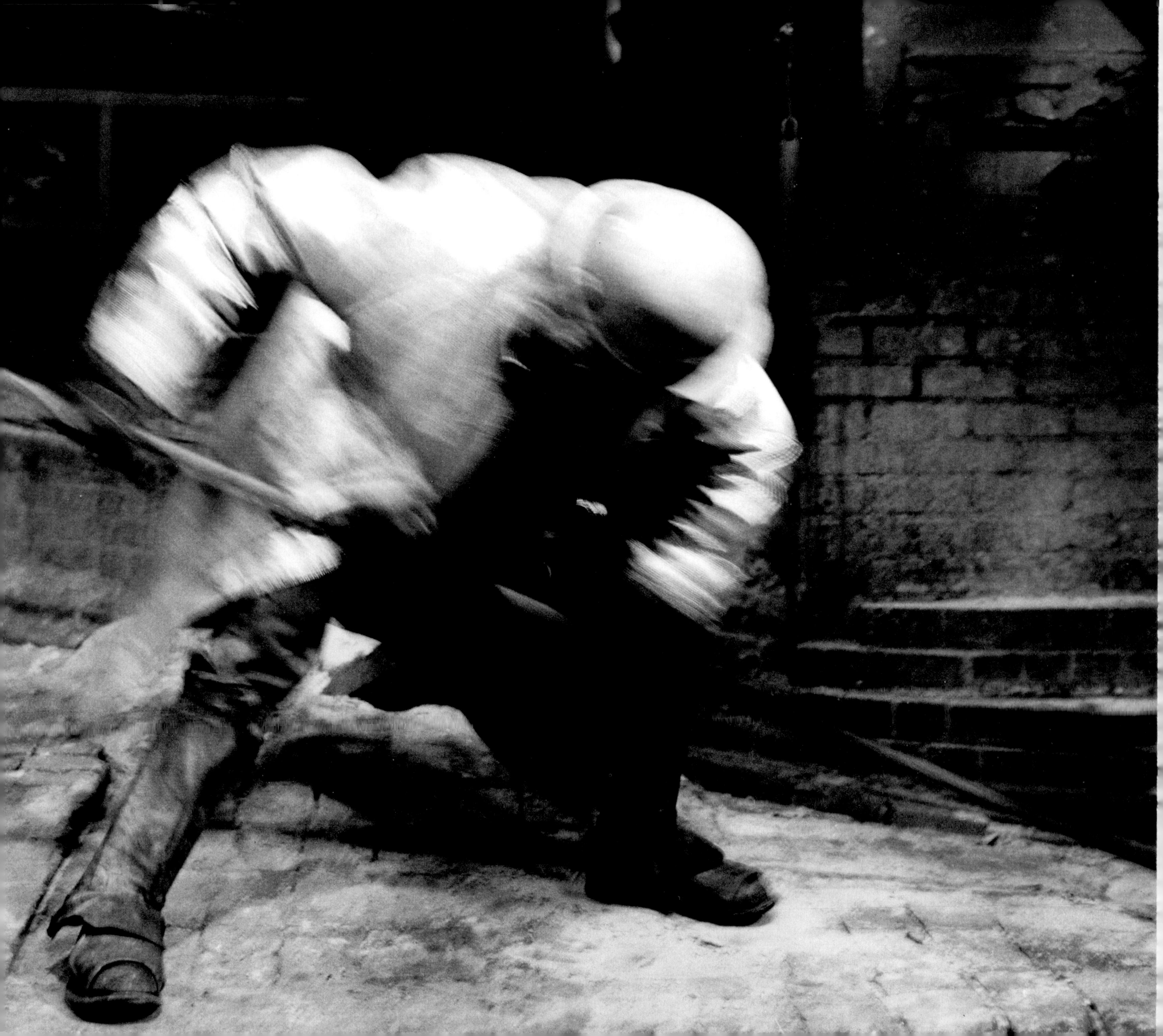

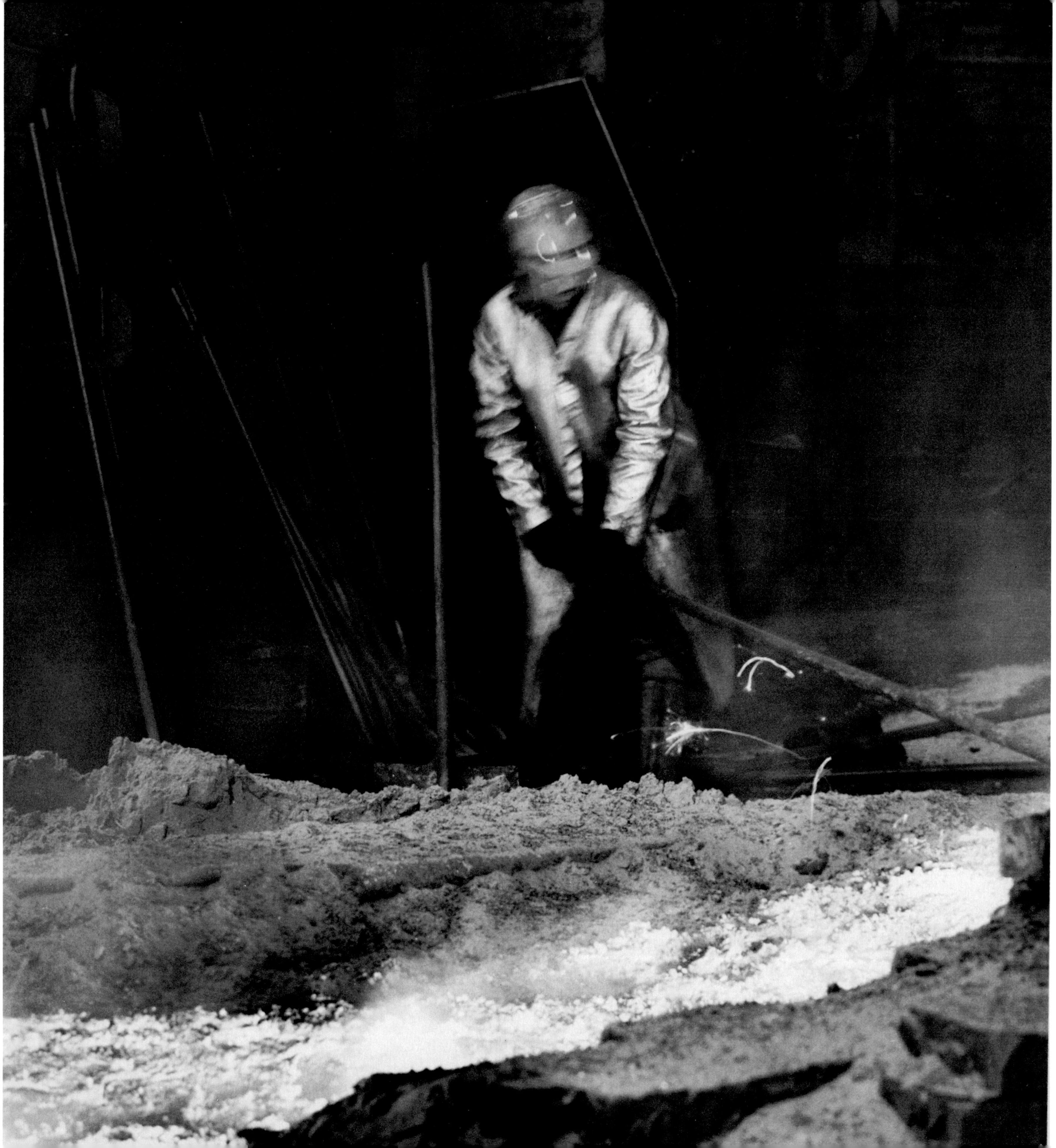

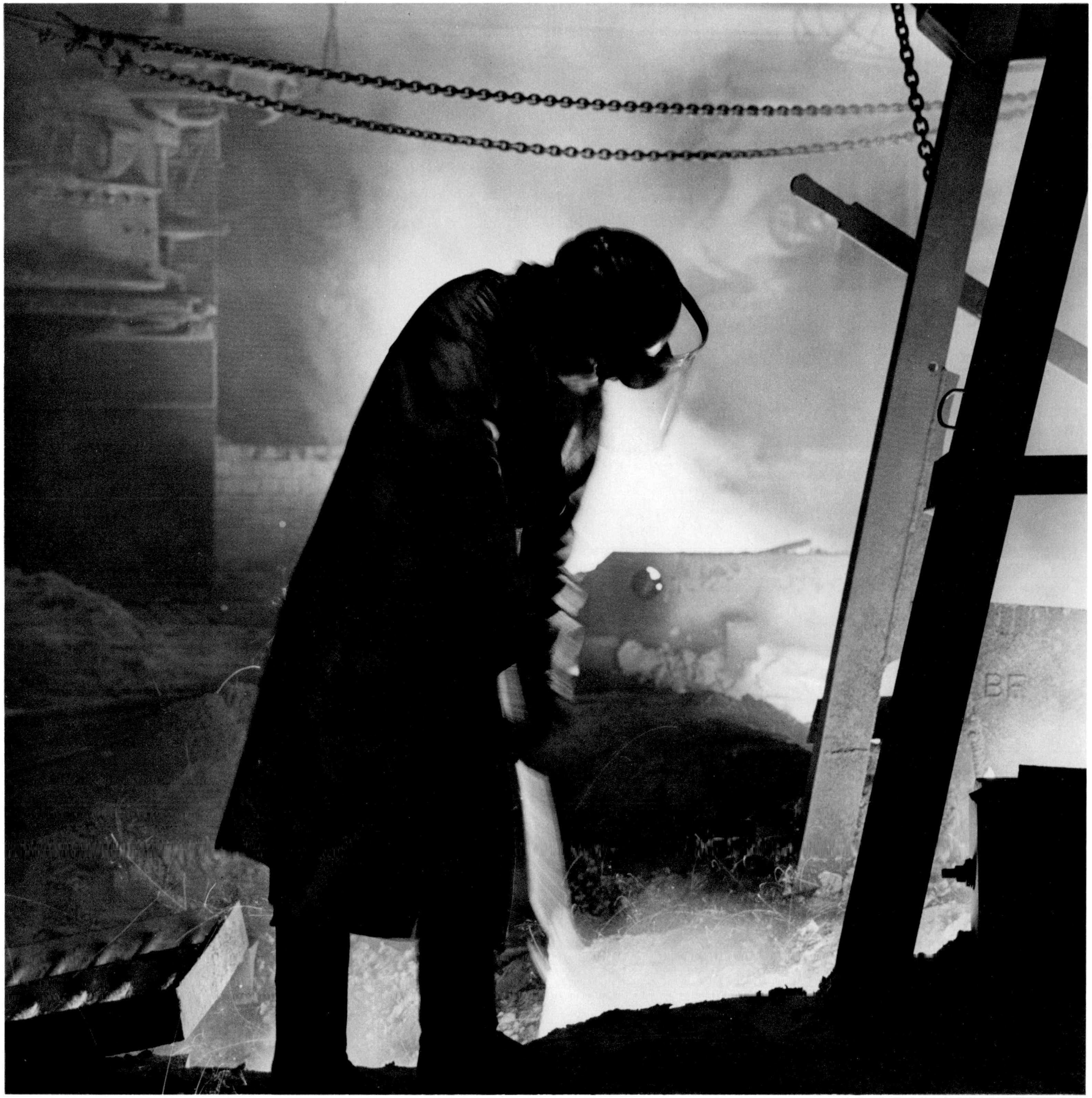

Casting.

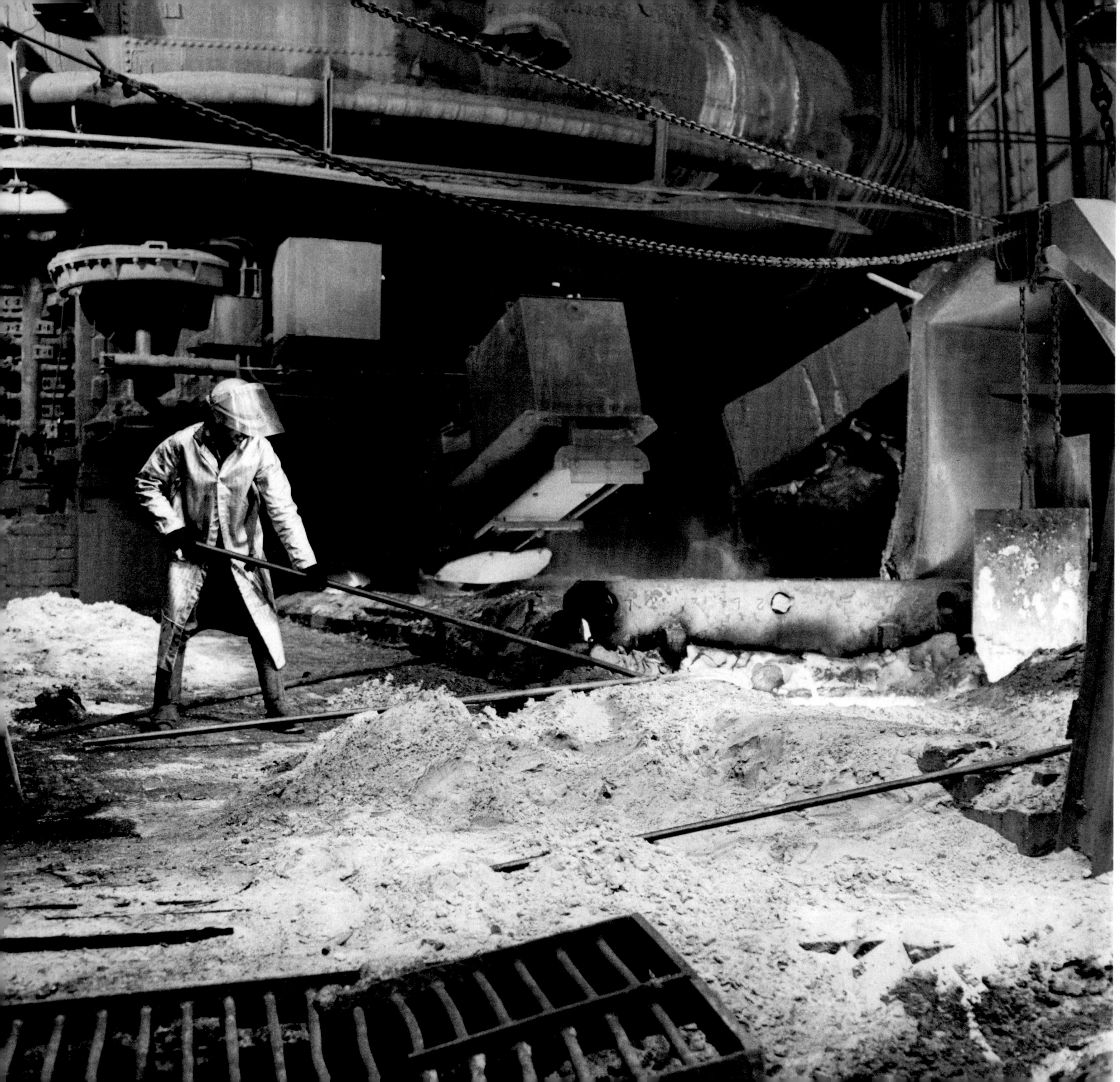

Casting.

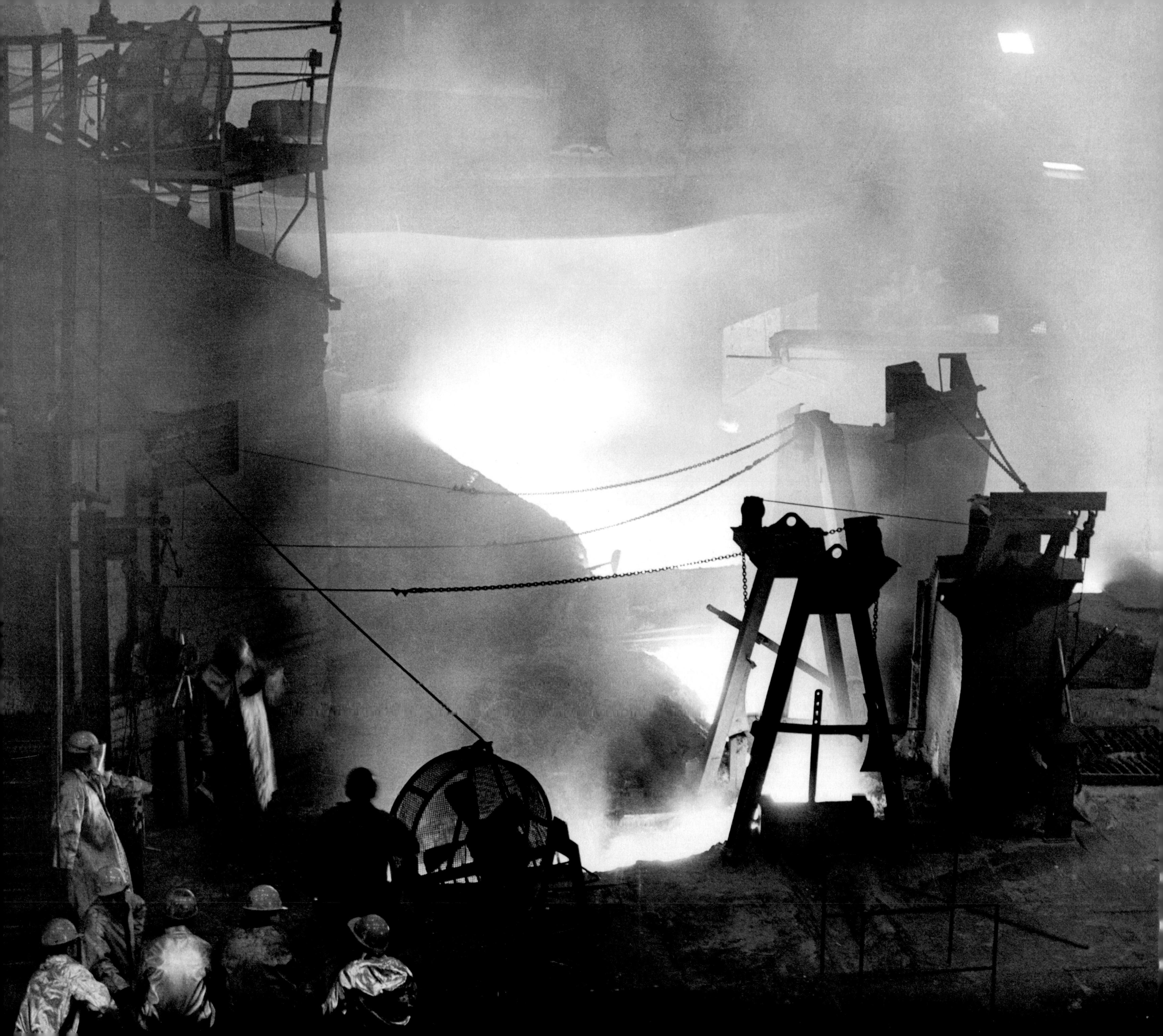

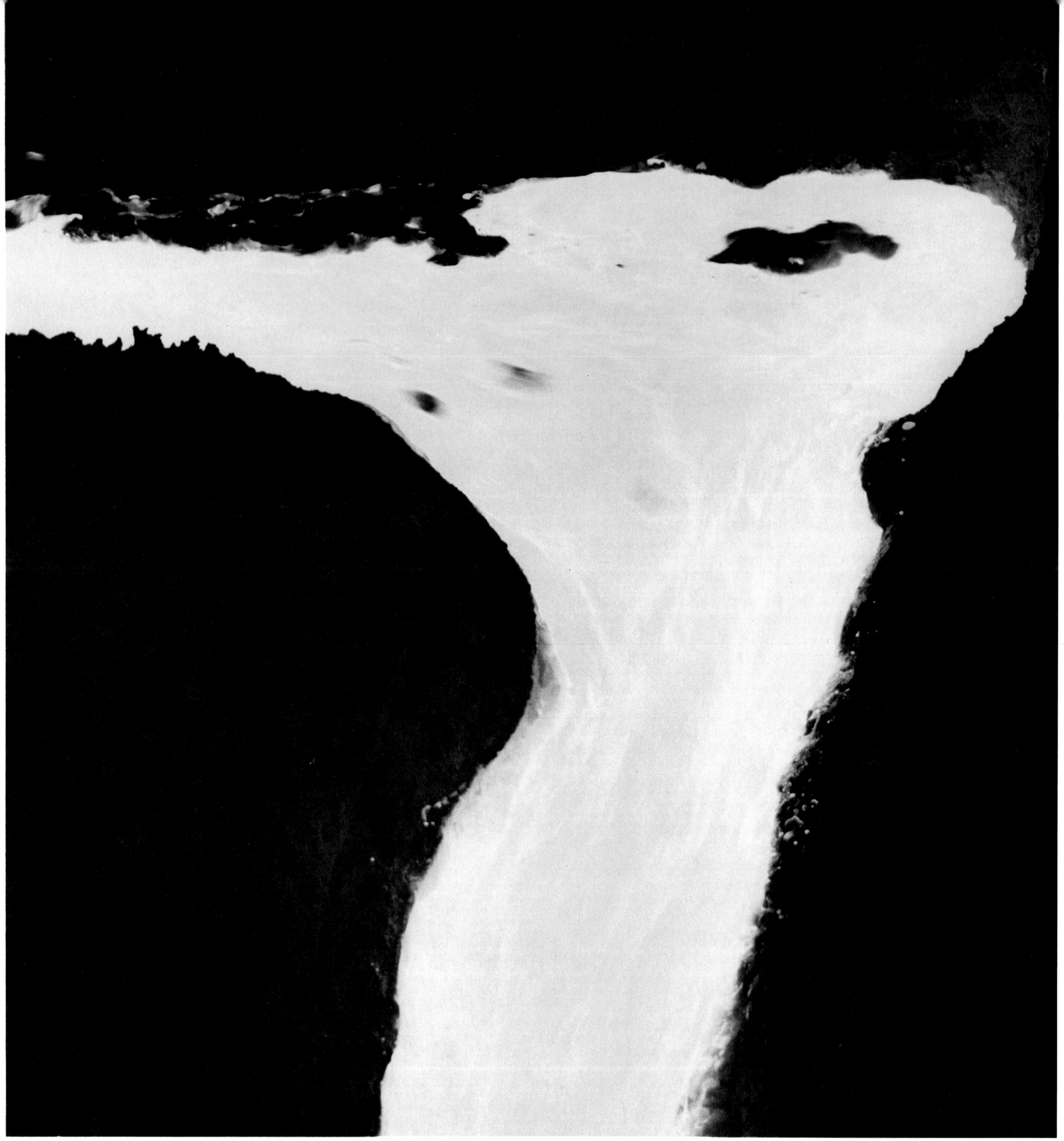

Molten iron in a runner.

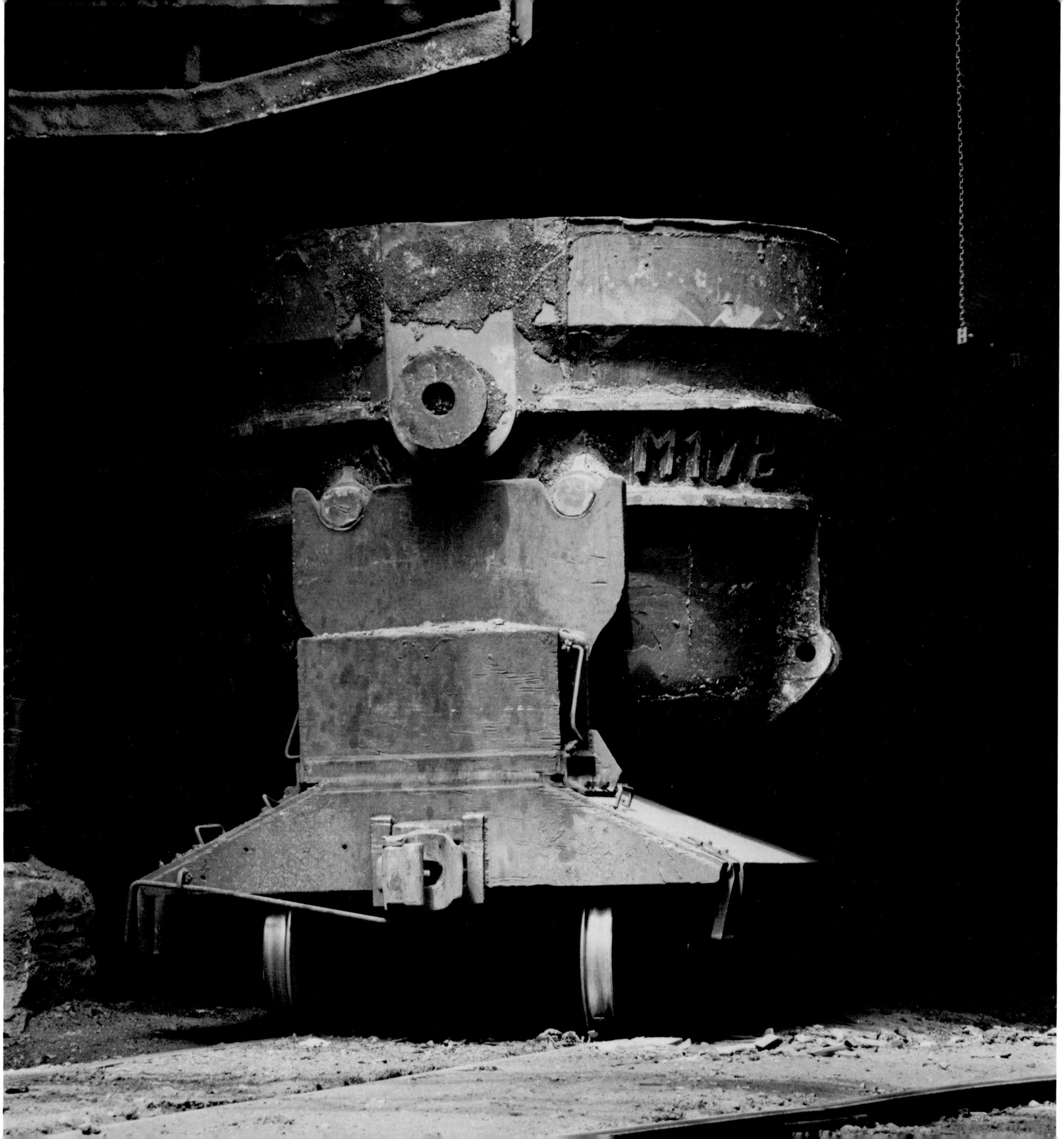

A slag pot.

View of the mill from the
top of a blast-furnace stove.

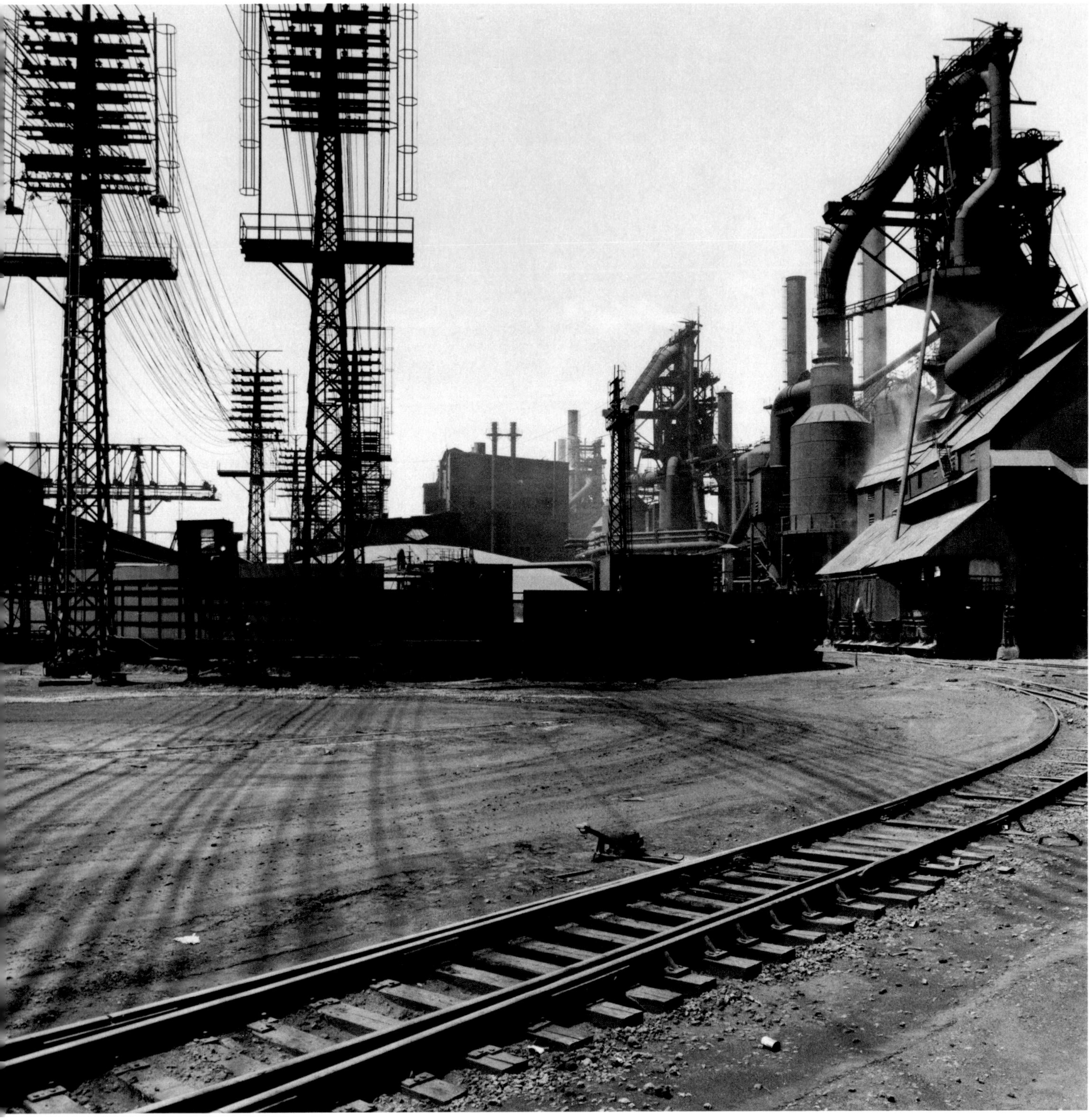

Blast furnaces.

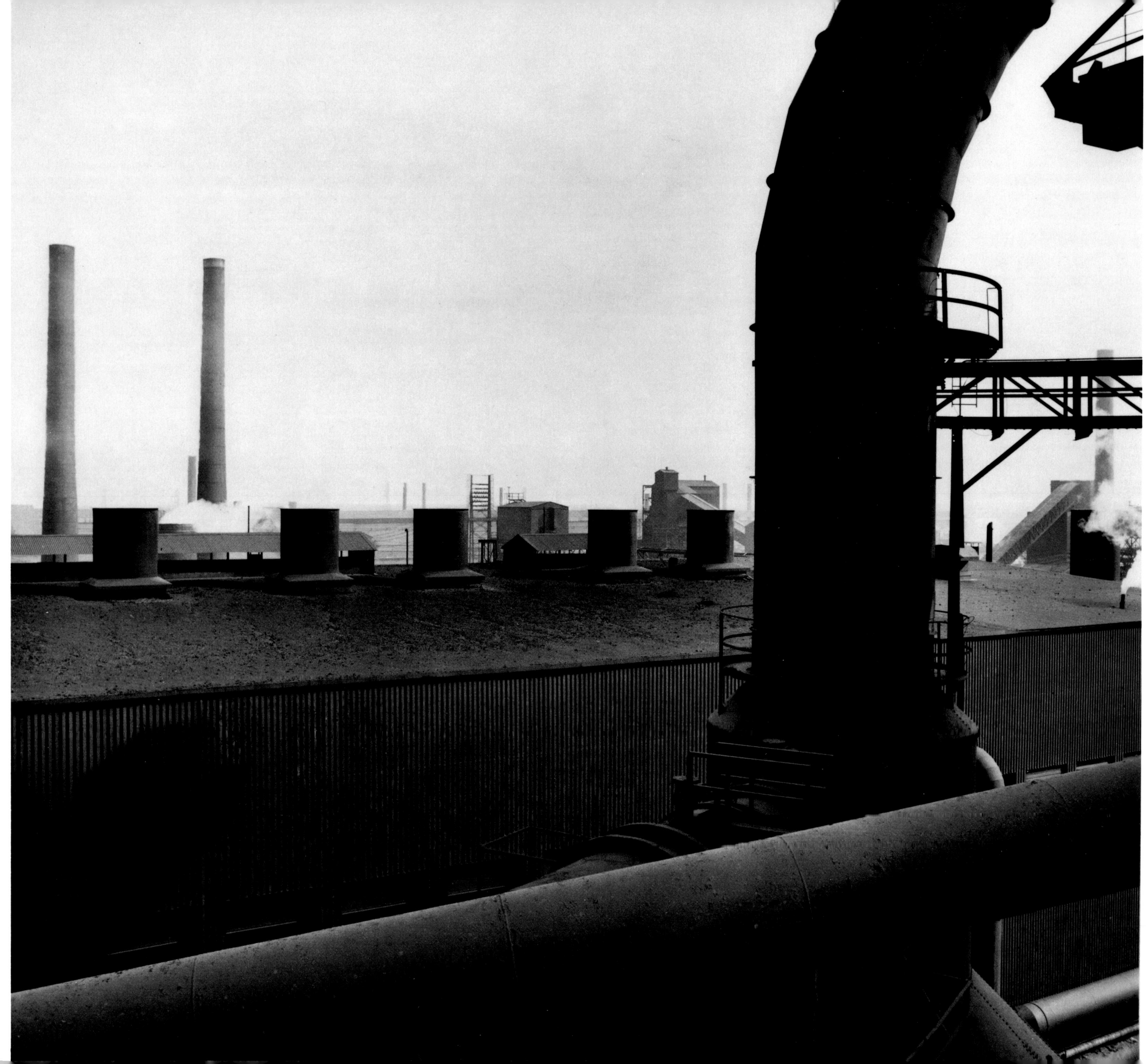

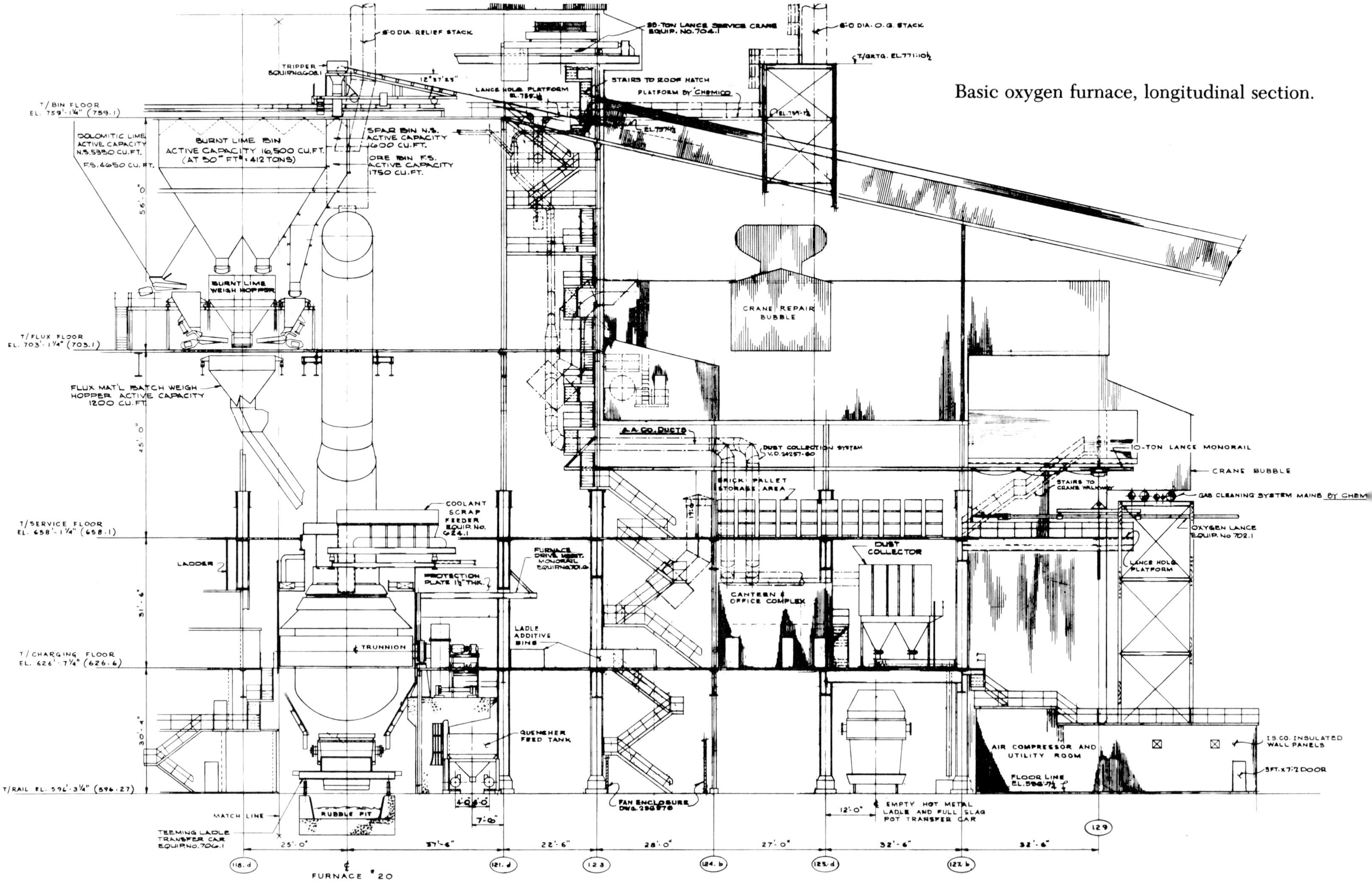

Basic oxygen furnace, longitudinal section.

The Making and Shaping of Steel

The critical difference between iron and steel is that steel contains less carbon than iron. Pig iron contains about 4 percent carbon; steel has only about .4 percent. The steel furnaces, then, are essentially iron refineries, their primary objective being to reduce the amount of carbon in the iron. Most steel is actually known as carbon steel, whose properties are determined by its carbon content. But the reduction of carbon is not the sole function of the furnace. Steel is more than just refined iron; the word is a generic term referring not to a single metal but to a singularly versatile substance. Steel is formulated to fit an almost infinite variety of purposes, and each batch, or "heat," as it is called, is blended to precise specifications for each customer's needs. The particular characteristics of each heat are derived from the addition of one element, the removal of another, and the subjection of the steel to different temperatures.

Despite the awesome aspect of steel-making machinery, which seems to require only the roughest measurements, the making of steel is a critical business, a highly developed science that involves minutely accurate specifications. The recipes for different heats of steel are so exact that if a chef were to use the same standards in the preparation of his meals, he would measure ingredients not by teaspoons or in terms of a "pinch," but rather by individual grains.

Today most steel is made in either the basic oxygen furnace (B.O.F.) or the electric furnace. The open-hearth (Siemens-Martin) furnace, which largely supplanted the Bessemer process as the primary method of producing steel early in this century, has now been virtually replaced by the B.O.F. as the most important method of steel-making. The latter is by far the most efficient and economical way yet devised to make large quantities of steel. In the B.O.F., three hundred tons of steel are produced in about forty-five minutes; the open hearth would take between five and eight hours to produce a similar quantity.

Molten pig iron is not the only "raw" material used in the steel furnace. All three methods of steel-making utilize scrap. Technically, scrap is not a raw material; it is recycled iron and steel, not just junk but carefully selected and graded material. In the industry, there are three kinds of scrap: home scrap, prompt industrial scrap, and dormant scrap. The first consists of bits and pieces of iron and steel that have been collected from within the mill itself. The second kind is leftover material returned to the mill by a customer. The third is scrap in the conventional sense—automobiles, cans, and anything else that has already been used. The importance of scrap looms ever larger on the horizon of a world faced with an ever-diminishing supply of natural resources. Some sixty million tons of scrap are remelted annually during years of peak production.

The operation of the basic oxygen furnace is very simple. Paradoxically, its principle of operation is not altogether different from that of the Bessemer method of making steel, invented more than one hundred and twenty-five years ago. In the basic oxygen furnace, oxygen that is blown onto a molten bath of iron and scrap combines with carbon and other undesirable elements—such as silicon and phosphorus—and chemically eliminates them, thus creating steel. The process takes place in a pear-shaped vessel lined with refractory material. The vessel is tilted forward to receive a metallic charge, consisting of one-quarter scrap and three-quarters molten metal. The scrap is charged first and is followed by the hot metal, which, when poured over the scrap, melts it.

The vessel is tilted back into place and covered with a hood before the "water-cooled oxygen lance" is lowered into it. Pure oxygen is then blown at supersonic speed over the top of the bath. A flux, in this case lime, is added. As the "blow" continues, the temperature rises to as much as 3,000 degrees F. Then, as the steel-making process nears its finale, the lance is retracted. The vessel is tilted forward once again to sample the carbon content of the "heat" and to take its temperature. When this procedure is completed, the furnace is tilted in the other direction and the steel is poured into a ladle waiting below.

At this point certain alloys are added to the heat. Each alloy imparts its own particular quality to the steel, such as ductility, tensile strength, or resistance to heat or corrosion. For instance, tungsten, with a melting point of 5,400 degrees F., is added to make the steel heat-resistant; chromium and nickel, to make stainless steel; molybdenum, to increase the steel's hardenability. There are a host of other elements that might be added, including vanadium, titanium, columbium, and zirconium. "Addition agents" such as cobalt and copper, which add other properties, are also mixed into the steel at this point. No steel is made without manganese, the scavenger which removes the oxygen that in turn had removed the carbon.

Pure oxygen is blown at an extremely high speed over the top of the bath in the basic oxygen furnace, and, as with everything else associated with the industry, it is consumed in vast quantities. In this sense it is as much a raw material as any. Oxygen, like taconite and coal, is processed before it is used. It is made in plants that quite literally distill the atmosphere, separating the component gases from one another.

Ever since the days of the Bessemer converter, steel-making has been synonymous with spectacular displays of pyrotechnics. This is not the case with the B.O.F., whose fireworks are hidden. Its most prominent structural aspect is its pollution controls. More space is devoted to the cleaning of the gases and collecting of the dirt and dust produced by the process than to the actual making of steel.

Today, fire and brimstone belong only to the open hearth. Before the advent of the B.O.F., the open hearth was *the* method of making steel. In its heyday it produced as much as 84 percent of America's tonnage. But time is running out for the open hearth: it is a slow process, and in a world ruled by the dictates of the uncompromising equation "time equals money," slowness is out. The open-hearth method is not without some advantages, however. It provides a greater degree of quality control than the Bessemer process and, until very recently, the B.O.F. It is also designed to use large quantities of scrap metal, which was an innovation at the time of its introduction.

The open hearth is a very different operation from its successor. It is so named because limestone, scrap, and metal are charged (in that order) directly onto a shallow hearth. There they are exposed—open to—the sweep of the flames that pass back and forth, alternating from either end of the furnace.

The furnace is charged on one side from a cavernous

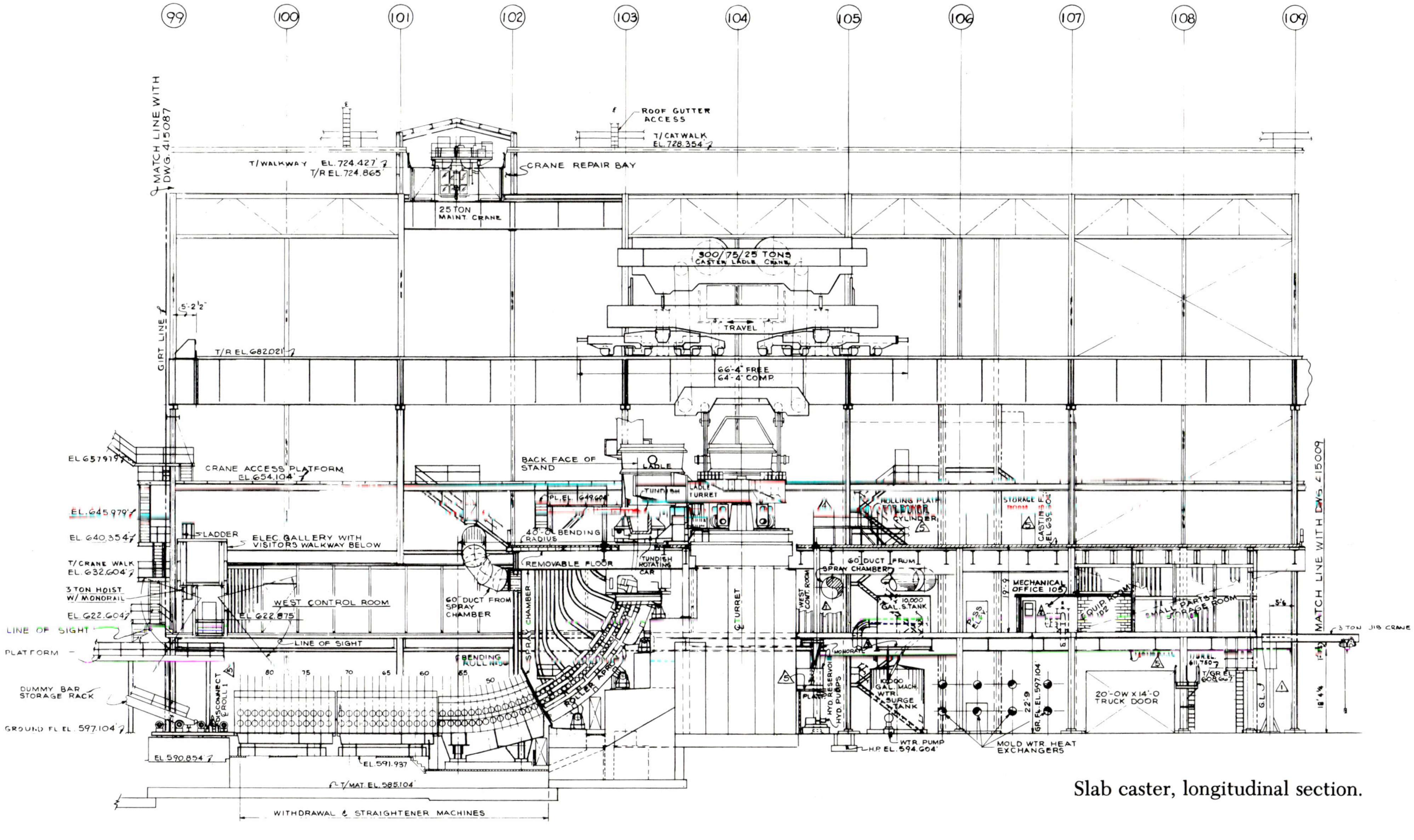

Slab caster, longitudinal section.

space known as the charging floor. Limestone and scrap metal are inserted first by means of devices called charging machines. When the initial materials have partially melted, hot metal is poured in, with spectacular results. During the process, the heat of the flames causes the impurities to combine with the limestone to form slag. The furnace is fueled by natural gas and oil, which today are usually supplemented with pure oxygen in order to increase the temperature and speed up the procedure.

Everything about the open hearth is dramatic, even the way in which it is tapped. The tap hole is blown out by an explosive charge, creating a white-hot shower of molten steel. The typical open-hearth furnace is quite large, often ninety feet long and thirty wide. Shaped rather like a split-level house, it can have a capacity of as much as six hundred tons. An open-hearth shop will usually have several—and often a dozen or more—individual furnaces, all lined up and each one with its charge at a different stage in the process of steel-making.

The electric furnace is the specialist among steel-making facilities. Traditionally it has been used primarily to make small quantities of specialty steels—stainless steel and tool steels, for example. These products call for a particularly precise degree of temperature control, which only the electric furnace can provide. More recently, electric furnaces have been designed with capacities similar to the basic oxygen furnaces, and they can produce relatively large quantities of carbon steel.

Electric furnaces are unique in another way: they do not use hot metal, for their charge derives entirely from scrap. They can therefore be used to make steel in situations where hot metal is not available, such as in foundries and other types of shops that do not necessarily need to be adjacent to a blast furnace.

The electric furnace is a shallow cylindrical vessel. It is charged through the top by swinging the roof to one side. After charging with scrap and lime, the roof is swung back and three electrodes are inserted from above. The power, ranging from twenty-five thousand to eighty thousand kilo-volt-amperes, is turned on. Heat so intense that it melts the scrap is created by electrical current arcing back and forth from electrodes to the charge. The process takes about four hours to produce a "heat."

Finishing: Intermediate Stage

Straight from the steel furnaces in its molten state, raw steel is a useless commodity. A small amount of raw steel will assume its final solid state by being cast directly in its molten condition. But the greater amount by far will be worked—wrought—into its finished form. This feat is accomplished by three basic methods: forging, extruding, and rolling. Of the three, rolling is the most widely used process. Most steel assumes its ultimate shape in the rolling mill. All steel, save that which is cast, passes through an intermediate semifinished state—a sort of larval phase—en route to the rolling mill.

Traditionally, steel assumes its solid condition for the first time in the form of an ingot, an oblong block of metal weighing several tons. Ingots are formed by pouring—"teeming"—molten steel from a ladle into cast-iron ingot molds. Within these molds, the steel cools and solidifies from the outside toward the center.

Ingot-teeming is not a haphazard affair, not just a matter of pouring raw steel into a container where it will eventually turn into solid material. It is a critical process. The temperature of the steel as it is poured, the manner in which it is allowed to solidify, the shape of the mold itself, and even the material with which interior surfaces of the mold are coated will affect the properties of the steel and will play an important part in determining how the steel will ultimately be used. All these criteria are taken into account long before a single drop of steel has been teemed.

Once the steel has become solid, the ingot is "stripped"—the mold is lifted off with great tongs as a plunger holds the ingot in place on the "stool." The coating on the inside of the mold prevents the steel from sticking (in the same manner that a greased cake pan works). The stripped ingot, which is cooler on its surface than in its center, must be reheated to attain a uniform temperature before it can be worked with. This is accomplished in "soaking pits"—deep boxlike furnaces where the ingots are literally soaked in heat until they reach the proper temperature. When this point is reached, the majority of ingots are carried away on ingot "buggies" to the rolling mill.

Not all ingots are destined to be shaped in rolling mills. Some are taken directly to be forged—shaped by compressive force. The forging presses and drop hammers of the steel mills can exert as much as ten thousand tons of pressure—a far cry from the hammer-and-tongs methods employed by the village blacksmith. But the principle is exactly the same.

Although a relatively small amount of steel is forged today, it is often the best method of finishing steel for products that will be subjected to very high degrees of stress—turbine shafts for ships and generators or diesel-locomotive wheels, for example. An array of smaller products requiring the same treatment may also be "die-forged" from semifinished pieces of steel that have been rolled first. Forging and rolling not only shape the steel, they make it stronger. The extreme pressure exerted on it during these processes actually changes the metal's crystalline structure. In spite of its mass, an ingot is relatively weak. During the rolling, forging, and extruding processes, the crystals are broken down into ones that are smaller and more uniform—and therefore tougher.

The great majority of ingots are bound for the rolling mills. The first stop along the way will be in the blooming or slabbing mills, which are sometimes referred to as the primary mills. Here, the ingots are squeezed and pummeled—rolled—into blooms and slabs, semifinished forms from which all other shapes will eventually emerge. Blooms are enormous rectangular blocks that will become structural shapes—rails, girders, and I beams—which in turn will be used to erect bridges and skyscrapers. They are also used to make yet a third semifinished form, the billet. Billets are smaller, elongated blooms from which pipes, bars, rods for reinforcing concrete, and wire will evolve. Slabs, on the other hand, are wide, flat pieces of steel that are the basis of all flat-rolled products; they are the progenitors of automobiles and boilers, of ships and refrigerators.

Today not all steel is solidified into ingots. A recently de-

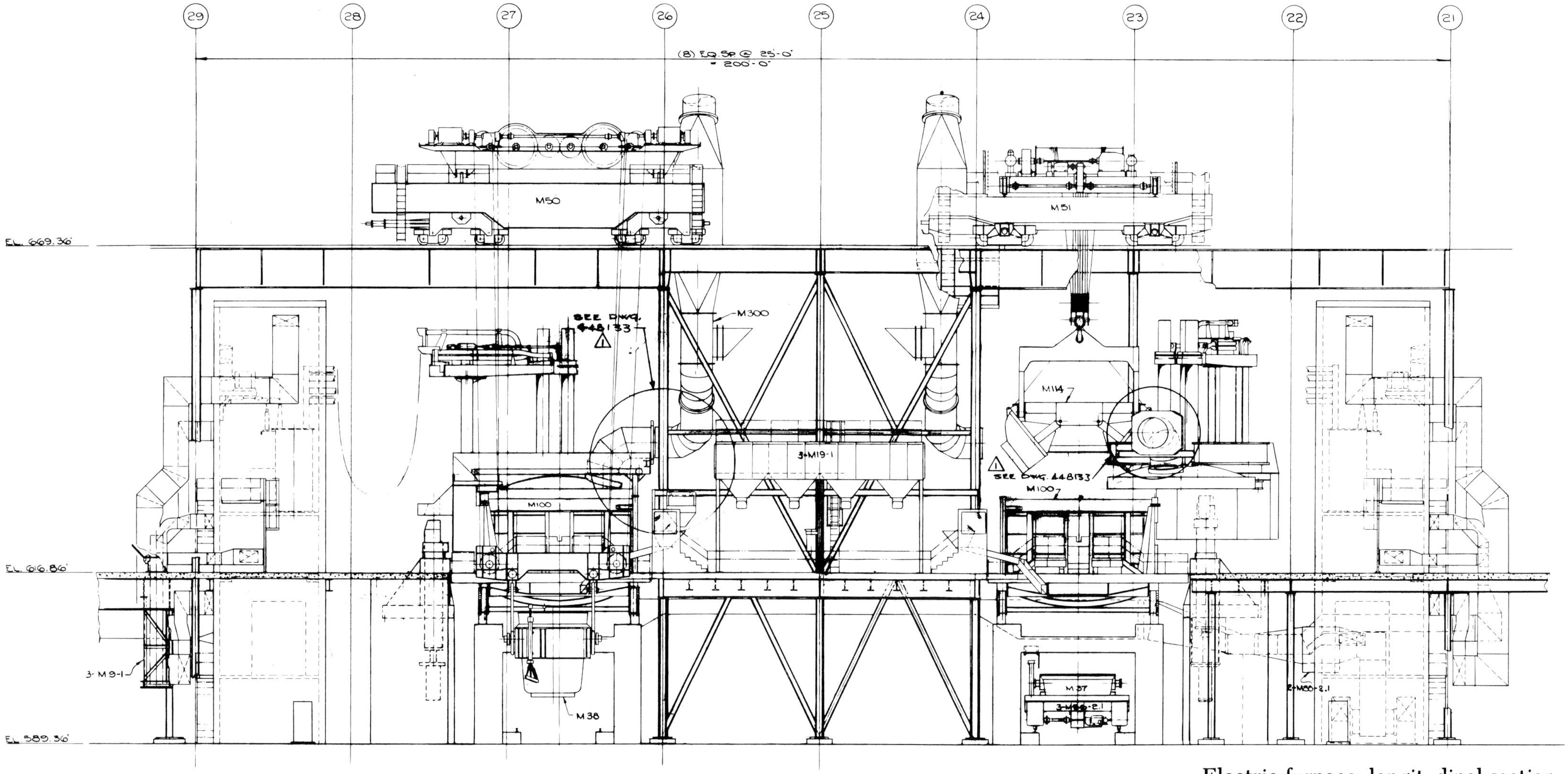

Electric furnace, longitudinal section.

veloped method known as continuous- or strand-casting has made it possible to eliminate the "larval" ingot stage altogether. The casting process produces instead a ribbon—a strand—of steel which when solid may be cut and trimmed into semifinished lengths. Blooms, billets, and slabs can then be produced without having to be rolled from ingots. This is a revolutionary step in the process of finishing steel, since it saves not only time and labor but material and energy as well.

The principles of continuous-casting are very simple; however, like most of the stages in the process of making steel, it is a very complex and exacting operation. Molten steel from the furnace is poured directly from the ladle into a shallow reservoir, the tundish. This bowl controls the flow of steel into the caster, a downcurving, water-cooled mold. Traveling deeper into the mold, the metal is cooled and begins to solidify as it is subjected to sprays of cold water. At the bottom, the steel becomes solid on a horizontal bed where it is cut and sent directly to the mills in which it will be rolled into finished shapes.

Rolling

All rolling mills perform essentially the same function: they mold steel into various shapes. Once again, the principle is very simple, the practice very complicated. Each particular shape and variety of finished steel requires one or more special mills to produce it.

Correctly, the term "rolling mill" refers to the actual apparatus itself. However, it has come to signify the place where the rolling operation is performed. A rolling mill consists of one or more sets of cylindrical steel rolls, which are placed one on top of or beside the other and which revolve in opposite directions. The rolls are arranged together in structures called "rolling stands." Pieces of red-hot steel ready to be rolled are passed from stand to stand or back and forth between stands on a bed of rollers known as the conveyor table. The space between the rolls themselves is always less than the thickness of the steel that passes through them; thus the metal is literally squeezed—kneaded like dough—into shape. The unbelievable pressure exerted on the metal as it passes between the rolls causes this to happen. The kind of shape depends on the type of roller. From a mechanical point of view, there are two basic kinds of rolling mills: reversing and continuous.

The reversing process is generally used in the primary mills—known also as the "bloomers" or the "slabbers"—and in the structural mills. The primary mill functions by passing the steel back and forth through the same rolling stand until the steel reaches a desired shape. The continuous mill operates by passing the steel in one direction through a set of stands. This process is used in the finishing mills, the bar mills, and for all flat-rolling—hot-strip and cold-strip—mills.

Given our consumer-oriented society, it is hardly surprising that 40 percent of all steel currently made in the United States is flat-rolled: it is from strip and sheet steel that automobile bodies, refrigerators, and washing machines are made. Flat-rolling mills are places that present extraordinarily graphic images, and for that reason most of the photographs of the rolling process were taken in an eighty-inch hot-strip mill.

Hot-strip mills are awesome places, often as long as three-quarters of a mile. In this distance—and in about four minutes' time—they can transform a red-hot twelve-ton piece of metal into a coil of millimeter-thin steel. The journey begins when the hot slab is dropped out of the reheating furnace onto the conveyor table. This alley of rollers will carry the metal from stand to stand. Throughout its trip down the length of the mill, each time the steel passes through a set of rolls it becomes thinner and longer and it moves faster and faster. By the time it reaches the coiler it may be as thin as one-tenth of an inch and as long as thirty-five hundred feet, and it may move as quickly as forty-five miles per hour. In the older mills, the meticulous job of adjusting the space between the rolls, which must be done each time a new order of steel is rolled, is entrusted to men. The entire operation of the newer mills is controlled by computers that have been fed with all the data necessary to custom-roll each slab of steel.

Hot-rolled steel may be considered a finished product in itself, and many million tons of it are sold each year. An ever greater percentage, however, is passed on to be further processed—cold-rolled, galvanized, annealed, or tempered. Whatever the next process may be, steel must first be "pickled" to remove the oxides from its surface. Pickling is usually accomplished by feeding the coils of steel through continuous pickling lines, where the steel is bathed in hot acid. Cold-rolling, so called because the coils are not preheated before they are rolled, subjects the steel to enormous pressures. This reduces its thickness even more and creates a finer surface, which may then be coated or painted. The pressures exerted within these mills are so great that it is possible to take hot-rolled steel the thickness of a half-dollar and reduce it to the thickness of two playing cards. At the same time, its length increases from three-quarters of a mile to more than two miles. Cold-rolling makes the steel even stronger, but it makes it harder as well. Hardness is not an asset; it makes the steel less workable. Therefore, most cold-rolled steel must be annealed to soften it again. Annealing takes place in specially constructed furnaces in which the coils are reheated and cooled under precisely controlled conditions. The process is so critical that it can take from four to five days to accomplish. Once annealed, cold-rolled steel—even that which is to be coated—may be considered a finished product.

Moreover, in this seemingly endless procession of finishing stages, steel that is not to be coated may be sent to the tempering mill, where its surface will be flattened and polished to an even finer degree.

Both hot-rolled and cold-rolled steel may also be galvanized—coated with hot zinc to protect it from that scourge of all ferrous metals, oxidation. This is generally effected by means of yet another continuous process: the "continuous hot-dip galvanizing line," where coils of steel are passed through a bath of molten zinc. The same procedure is used when coating steel with other nonferrous metals, such as aluminum, copper, or nickel.

Not all slabs are turned into coils. Many are rolled into plates. Without steel plates it has been said that the Industrial Revolution could not have occurred, for nothing else was strong enough to contain high-pressure steam within a boiler. Today about ten million tons of plate are used each year. Whenever great forces need to be contained—for ex-

Overleaf: Basic oxygen furnace and a lime kiln.

ample, in the hulls of ships, in nuclear reactors, in vessels made to contain liquid oxygen, and in gas pipelines—steel plate is still the only material strong enough to be used.

Among the many other types of rolling mills, the bar mills are quite important. The versatile bar can be rolled into almost every conceivable shape: round, hexagonal, square, rectangular, or oval, depending on its eventual use. From bars come a wide variety of products: small structural members, the raw material of hand tools, and rods. Rods are used to reinforce concrete or to form the basis of one of the steel industry's most important products—wire. Wire, it is said, has over a hundred thousand uses, everything from the cables of suspension bridges to paper clips. It is made by means of the third method of working steel, which involves drawing or extruding the rods through dies smaller in diameter than the rods themselves.

All of these intricate processes—the plants, the furnaces, and the mills—have turned raw material into finished products. Finished, however, only in the idiom of steel-making, for steel is itself a raw material. It seems almost as if by magic that rock is taken from the earth, melted down, refined, and then, in its molten state, solidified and shaped into forms that eventually emerge as automobiles, skyscrapers, boilers, and even scalpels. Magic or not, steel-making is an exacting and intricate scientific procedure, one that involves the highest degree of technical sophistication. A marvel of technology, steel-making is also the most spectacular of all industrial processes.

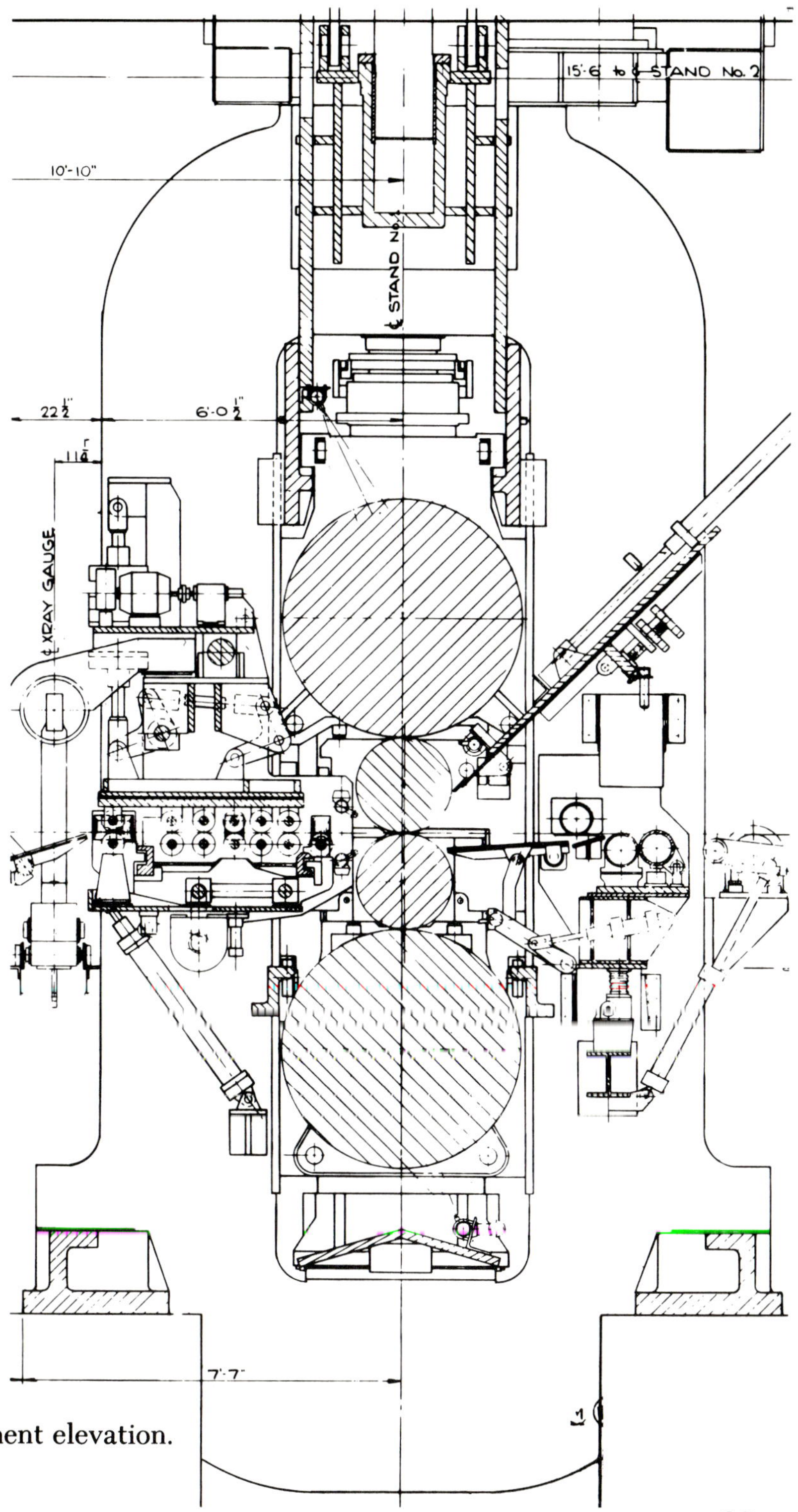

Detail of an 80-inch, 5-stand-tandem cold mill, general-arrangement elevation.

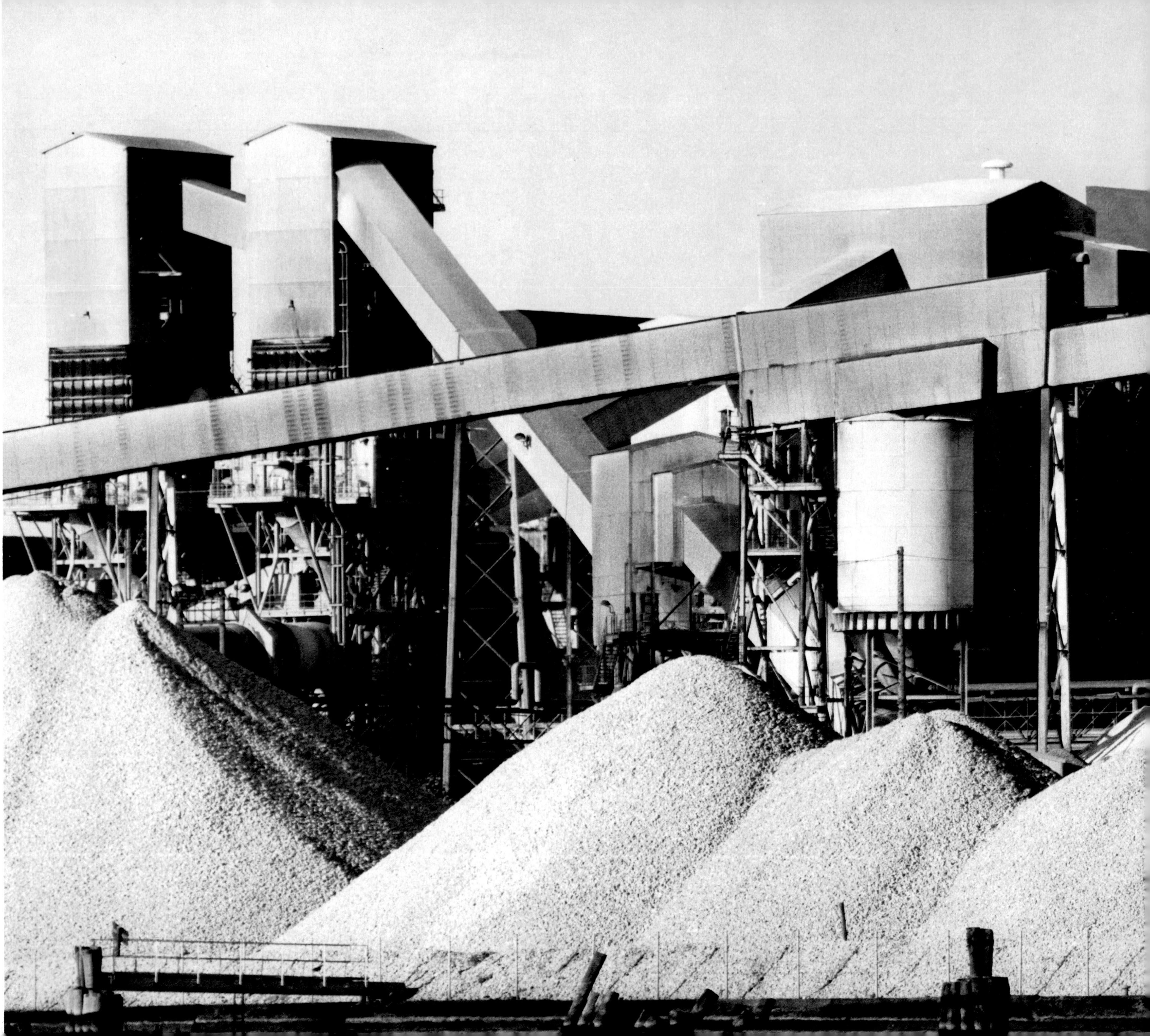

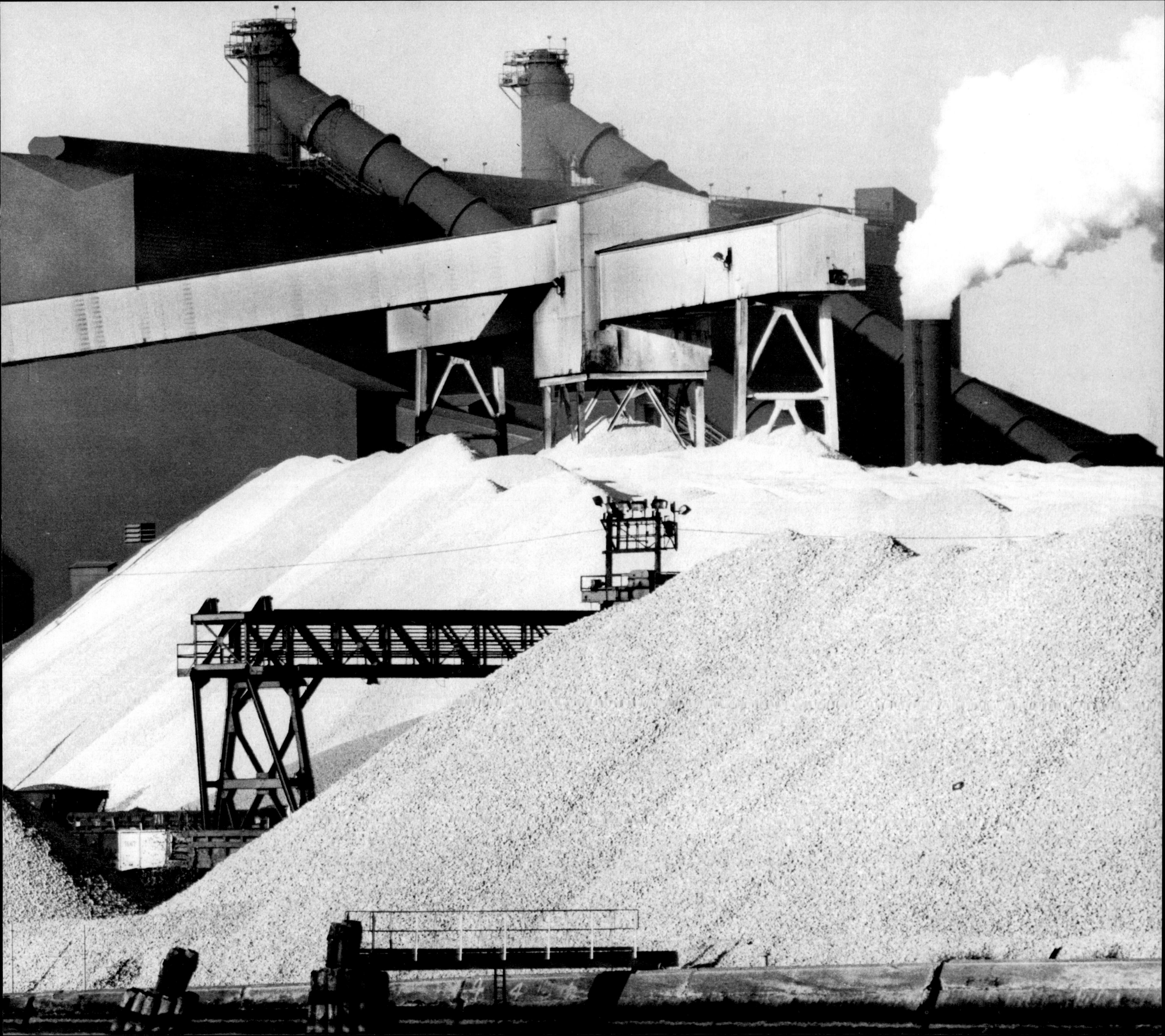

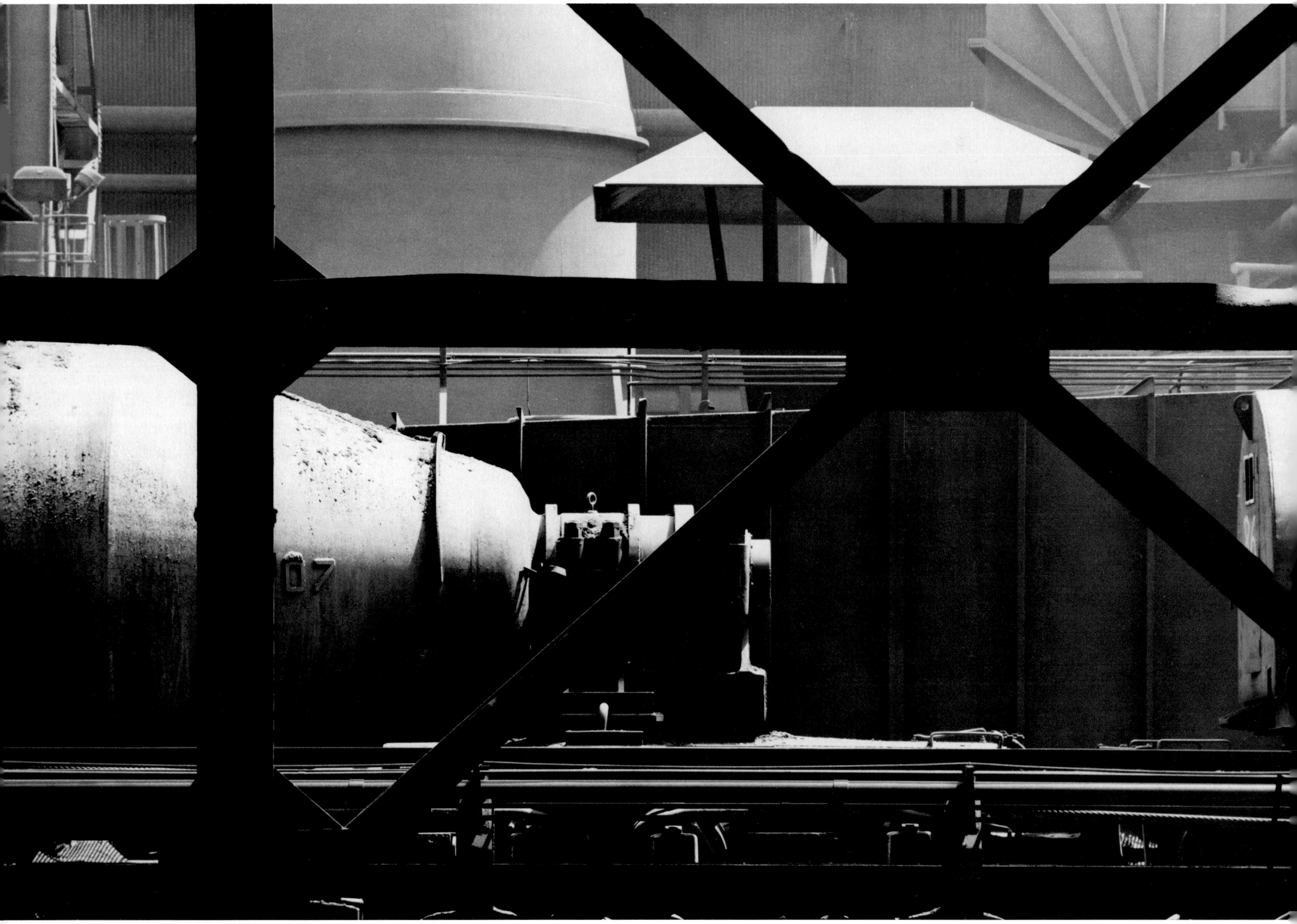
07

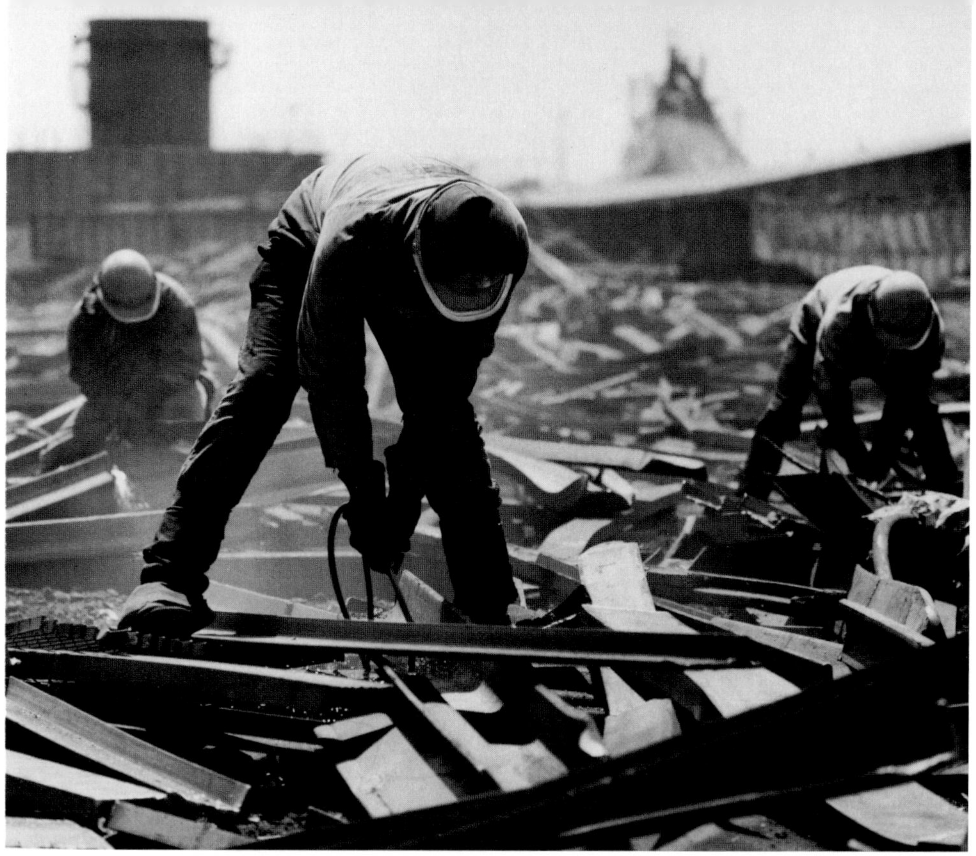

Cutting home scrap.

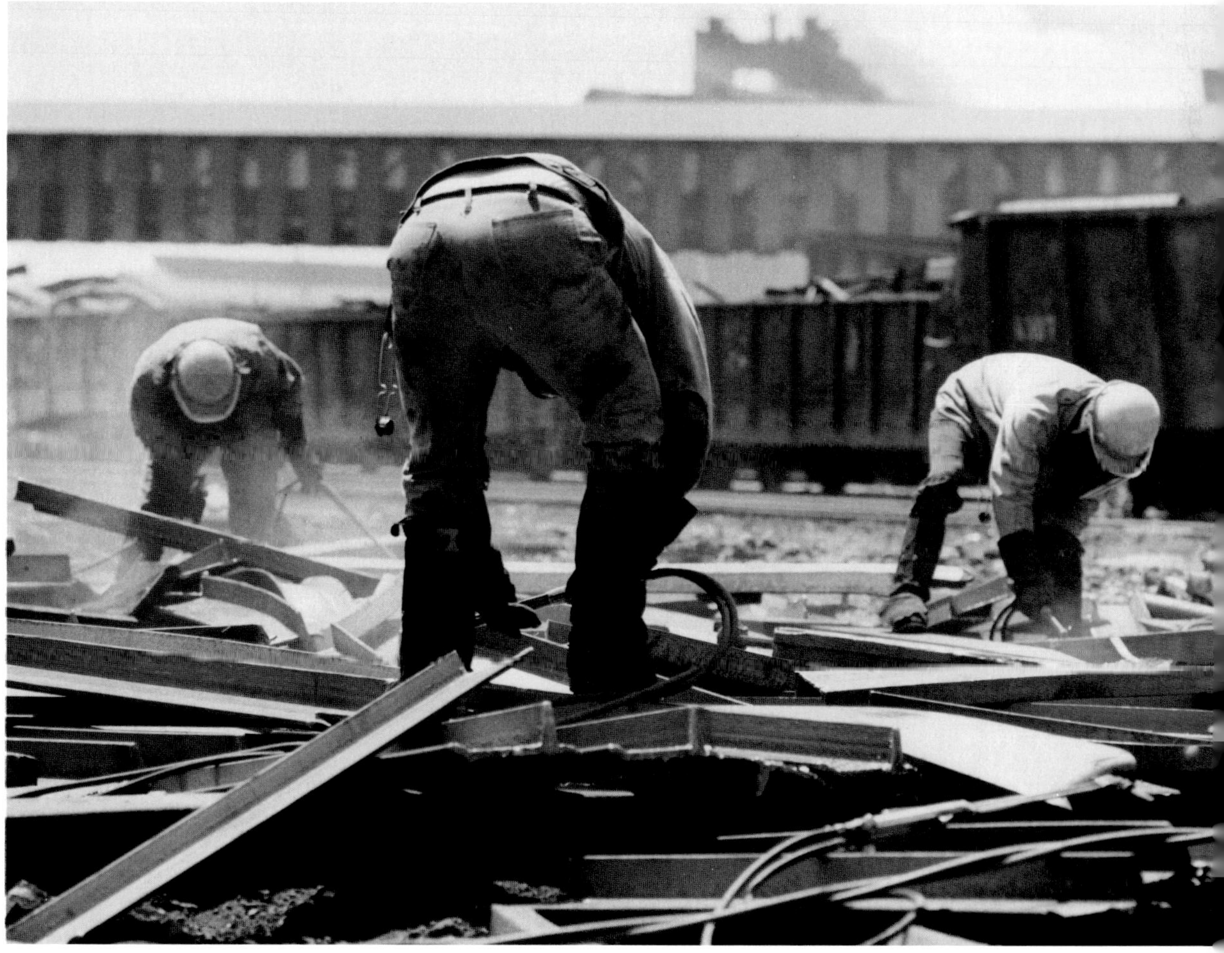

'ugh car.

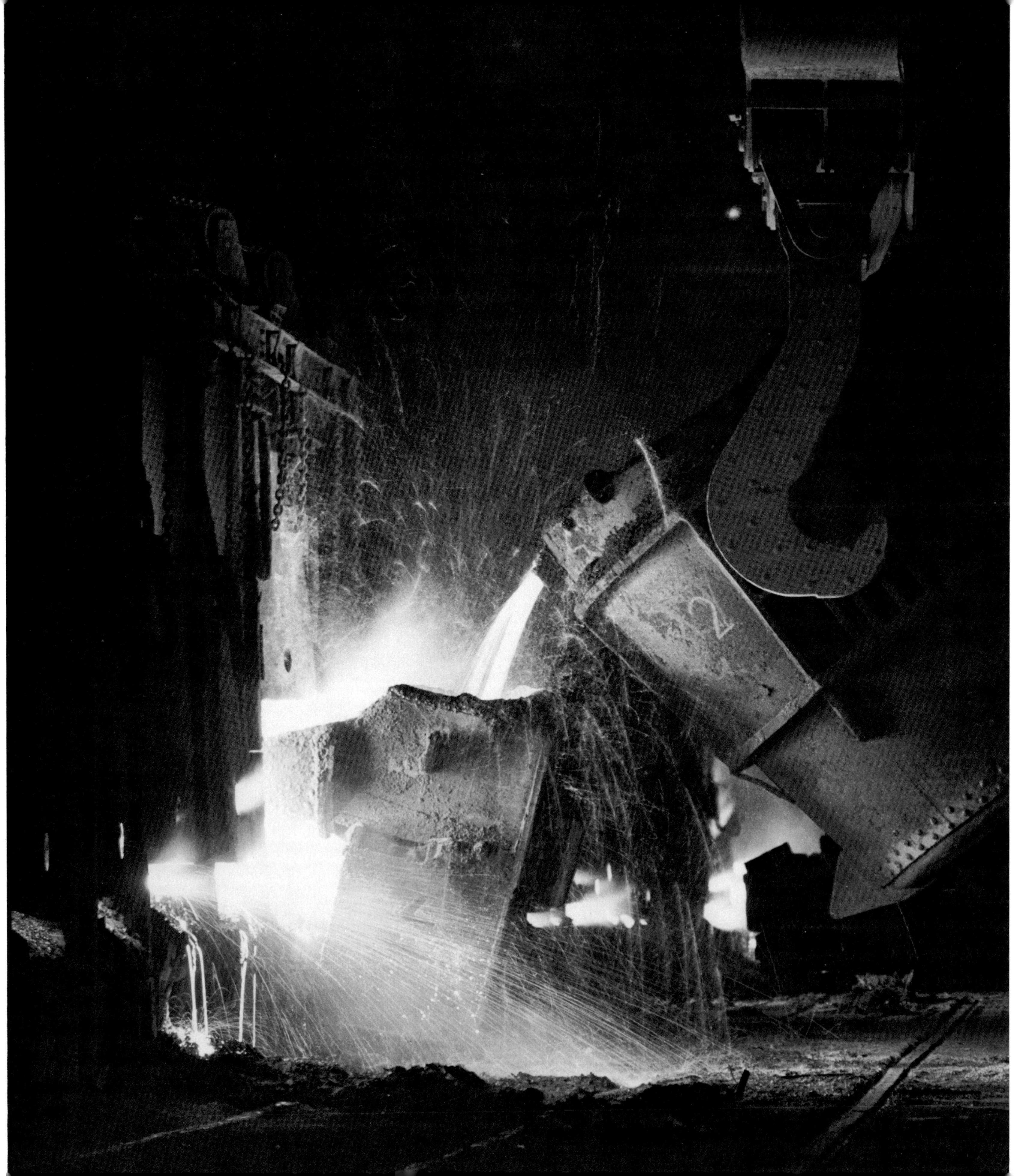

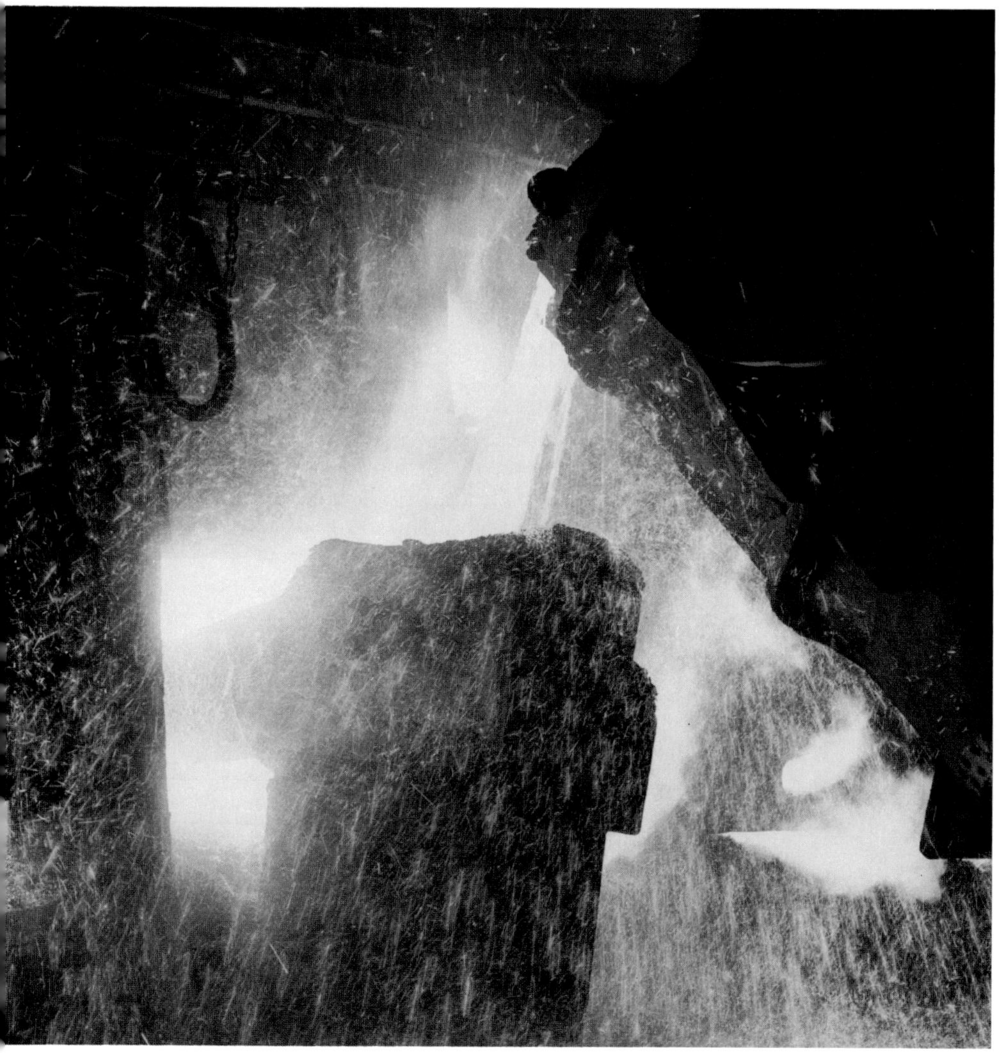

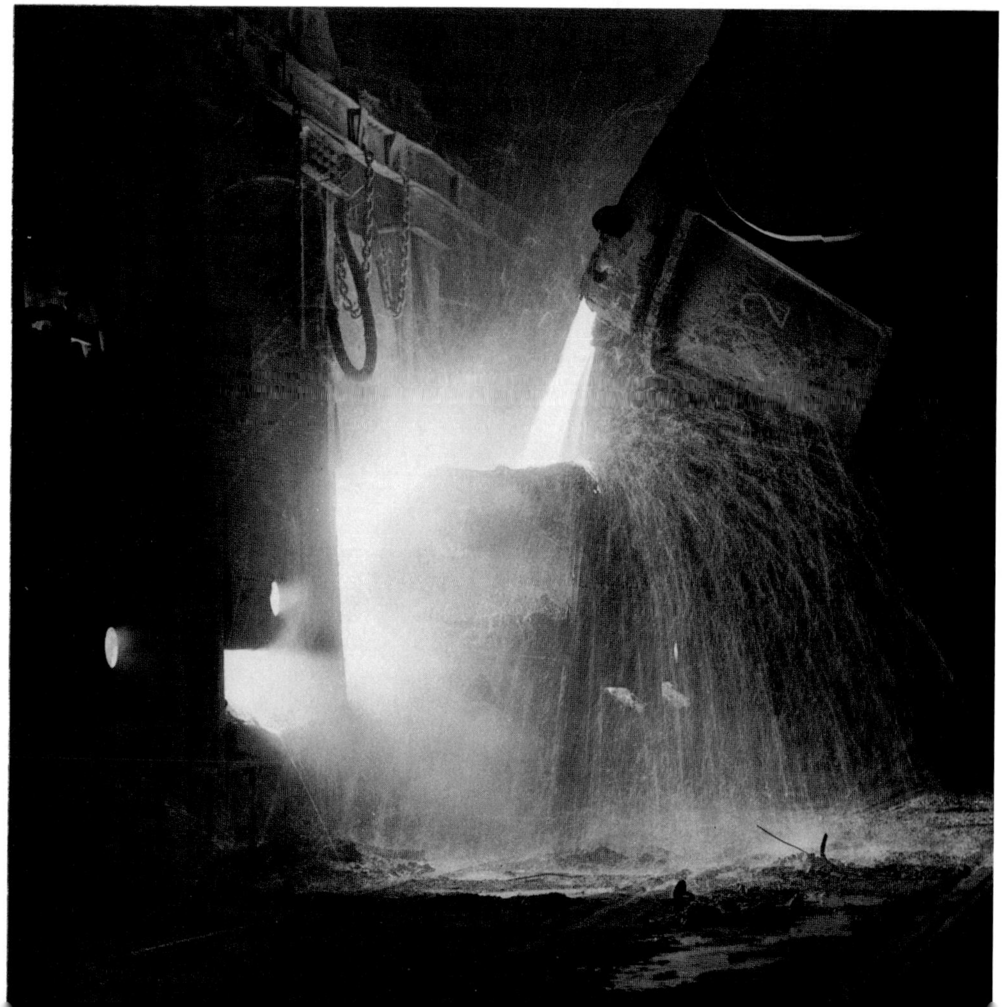

:harging hot metal in an open-hearth furnace.

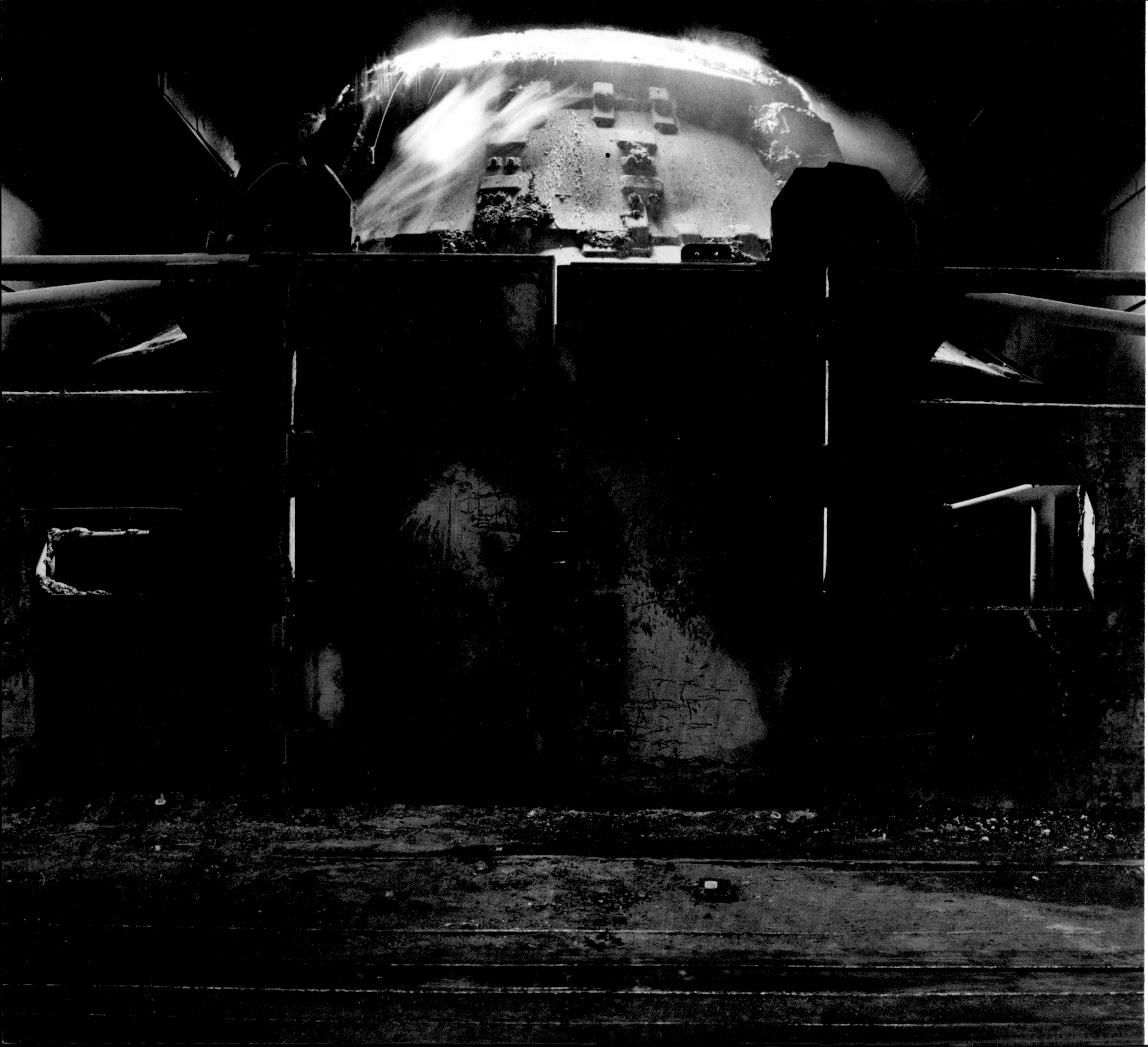

Basic oxygen furnace.

3lowing the bath.

Conducting one of a variety of tests.

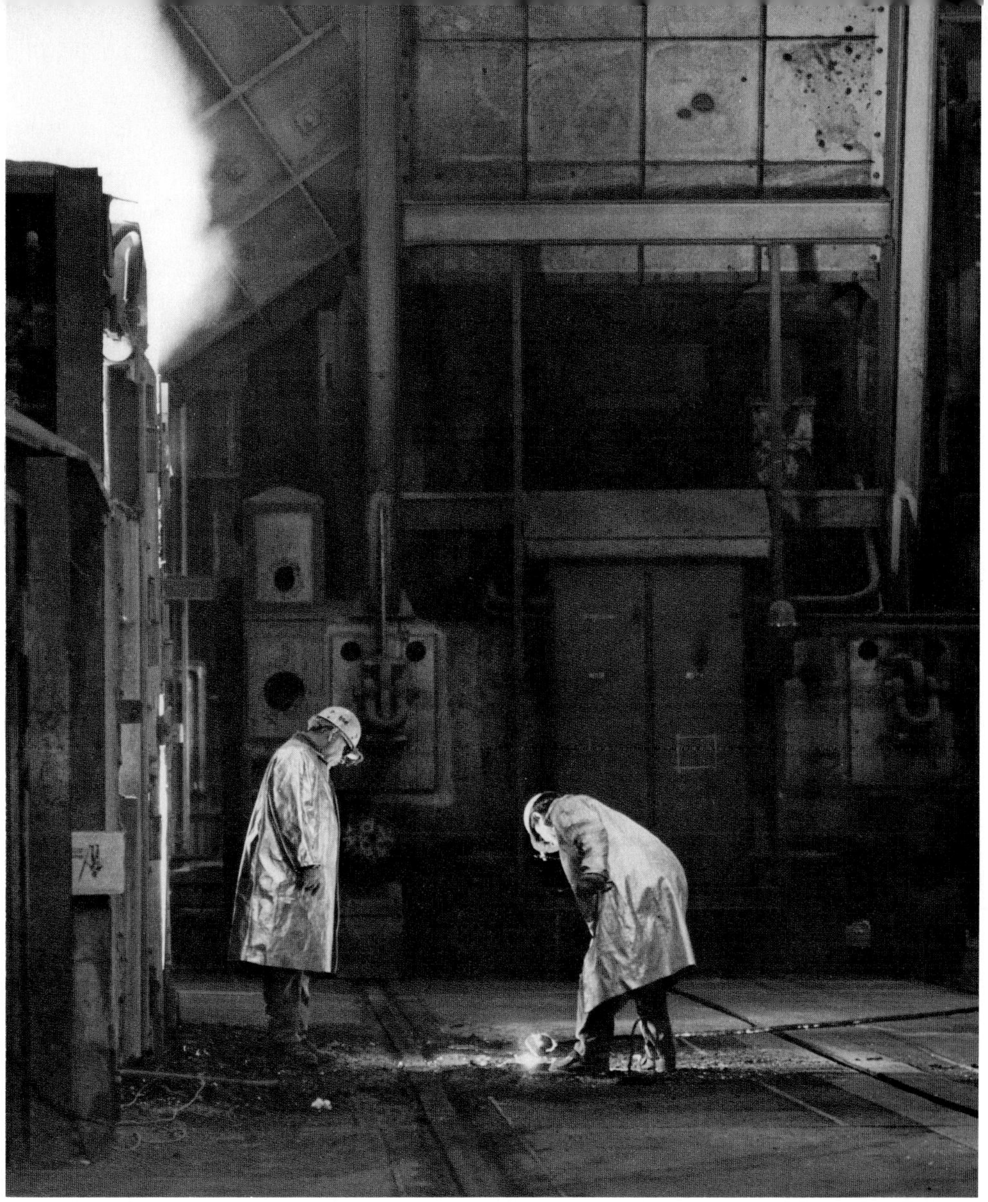

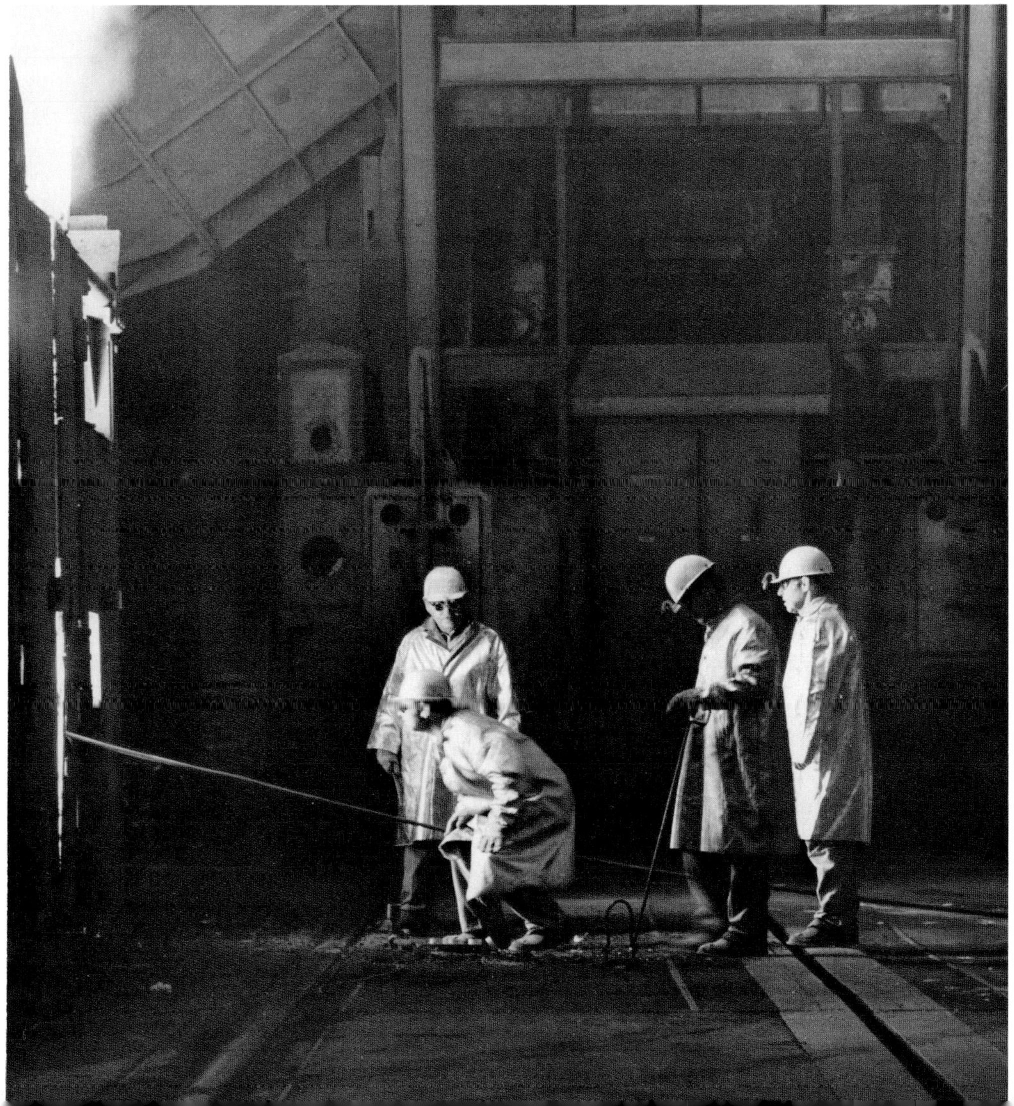

Taking the temperature of the bath.

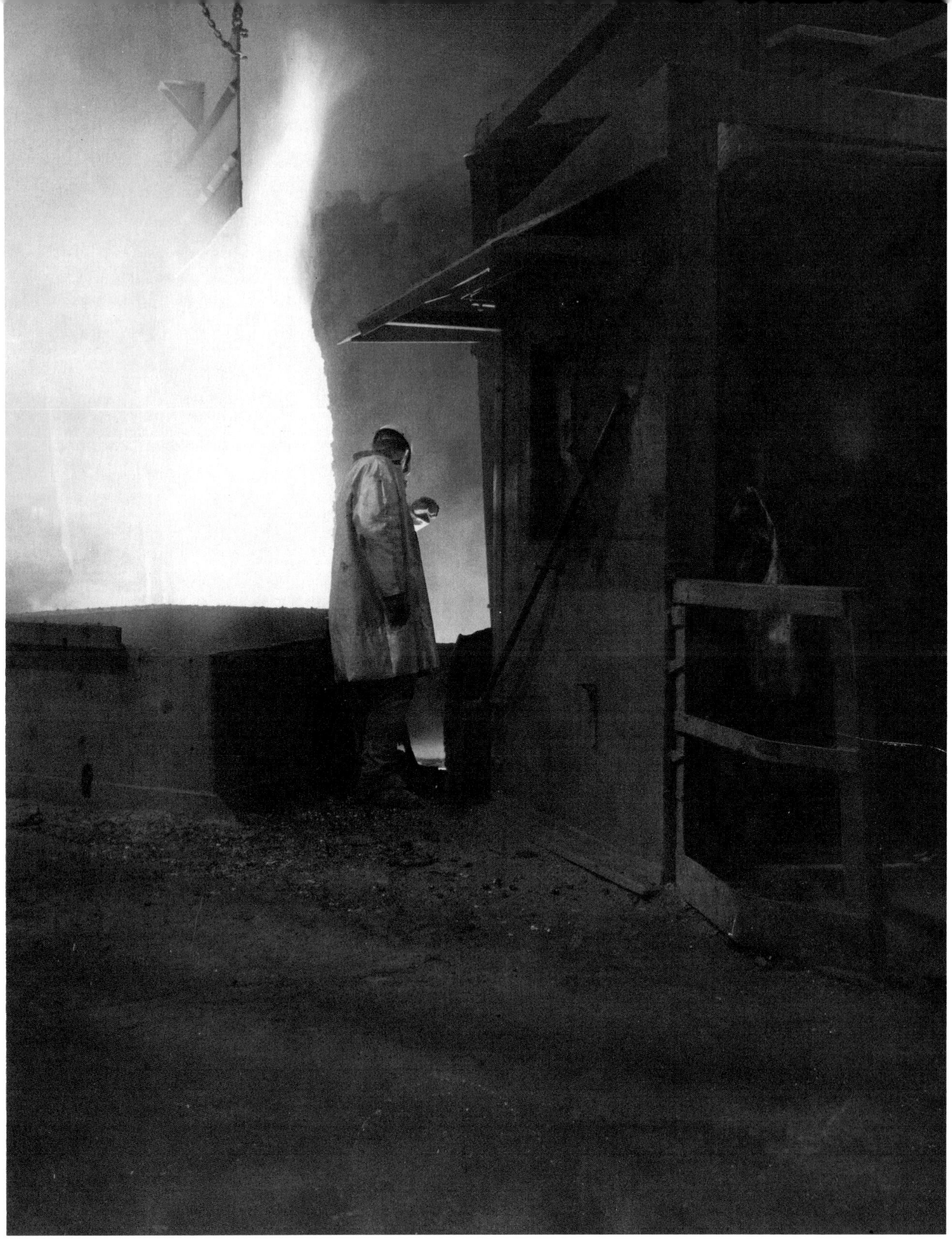

Tapping the heat in a B.O.F.

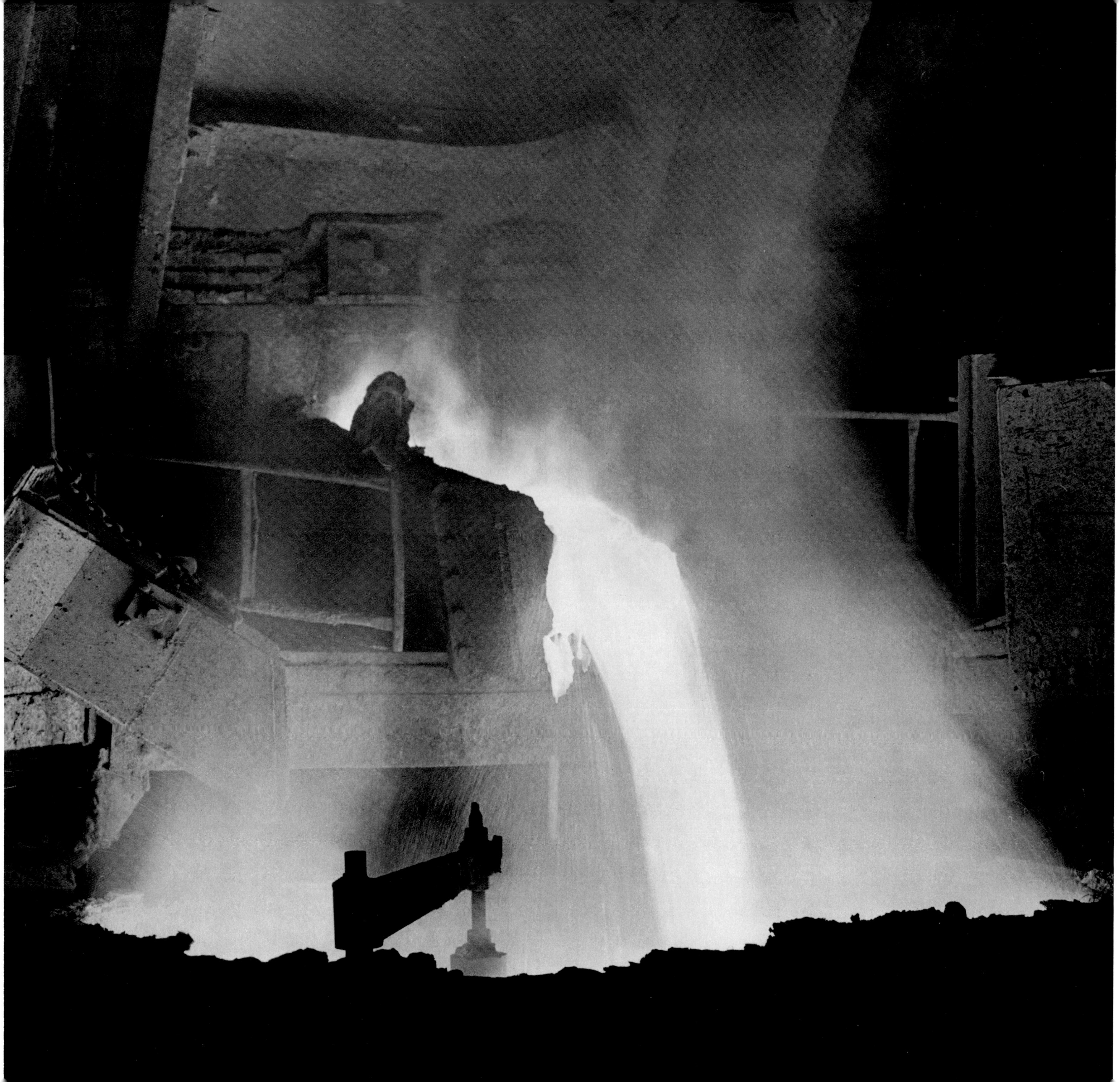

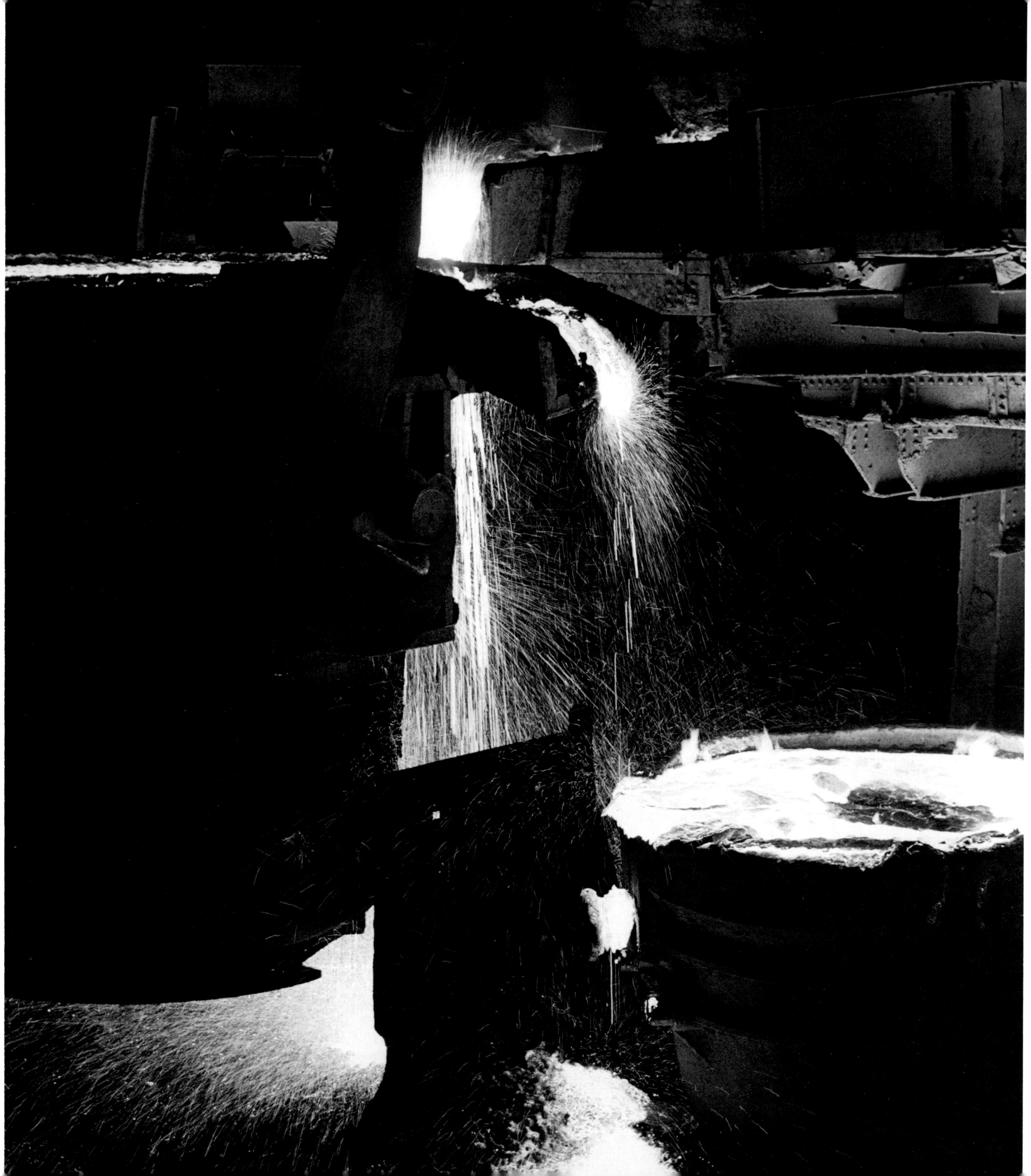

Opposite: ladle of molten steel and a slag thimble.

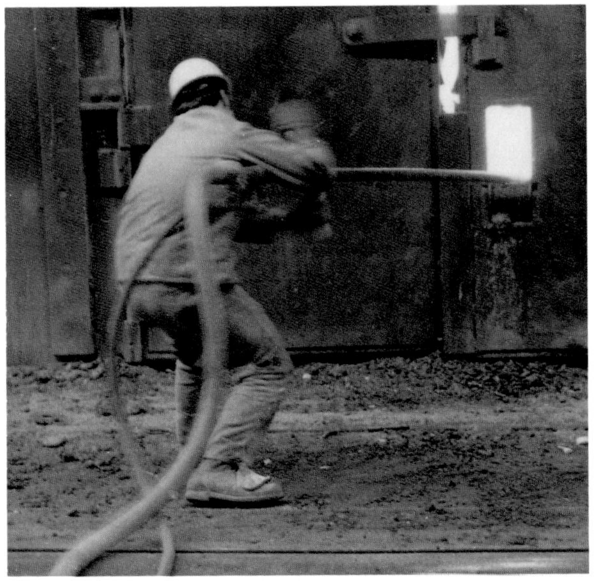

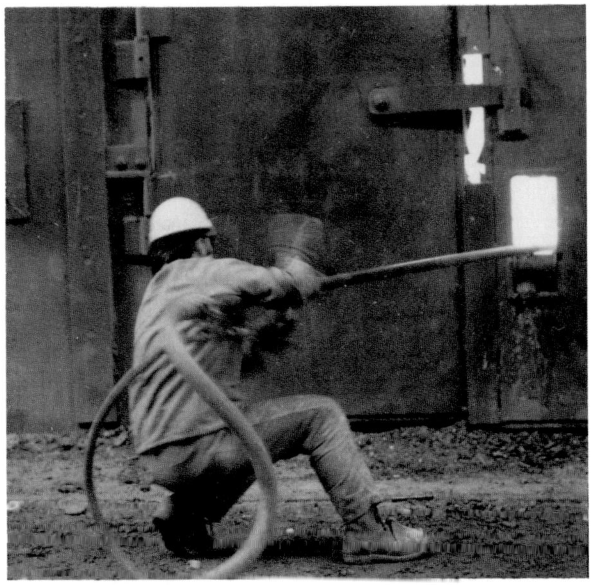

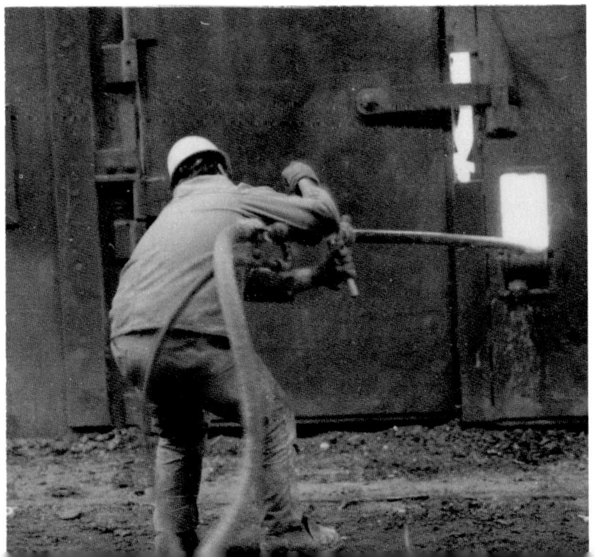

Spraying gunnite into a B.O.F. to patch refractory material within.

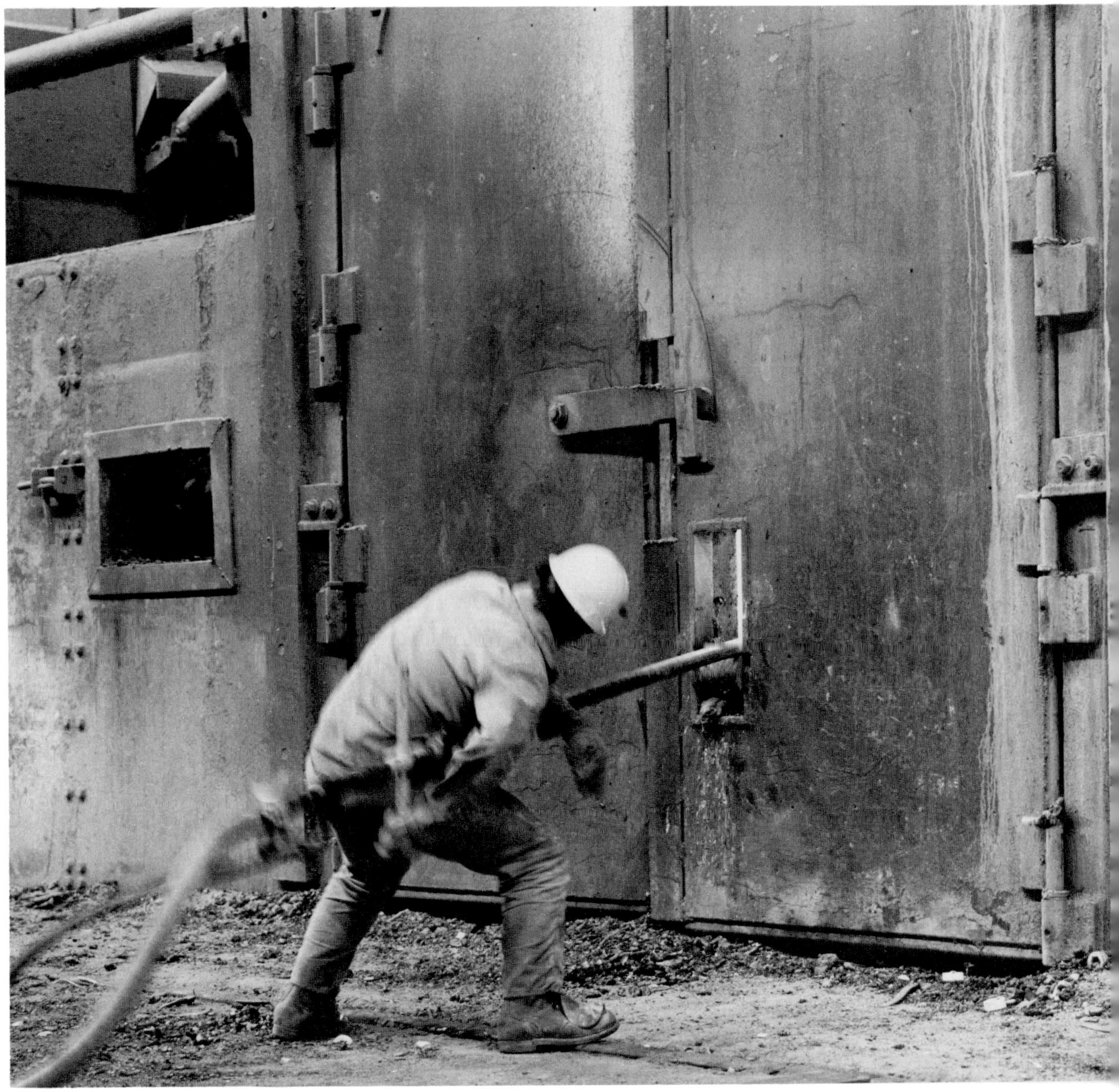

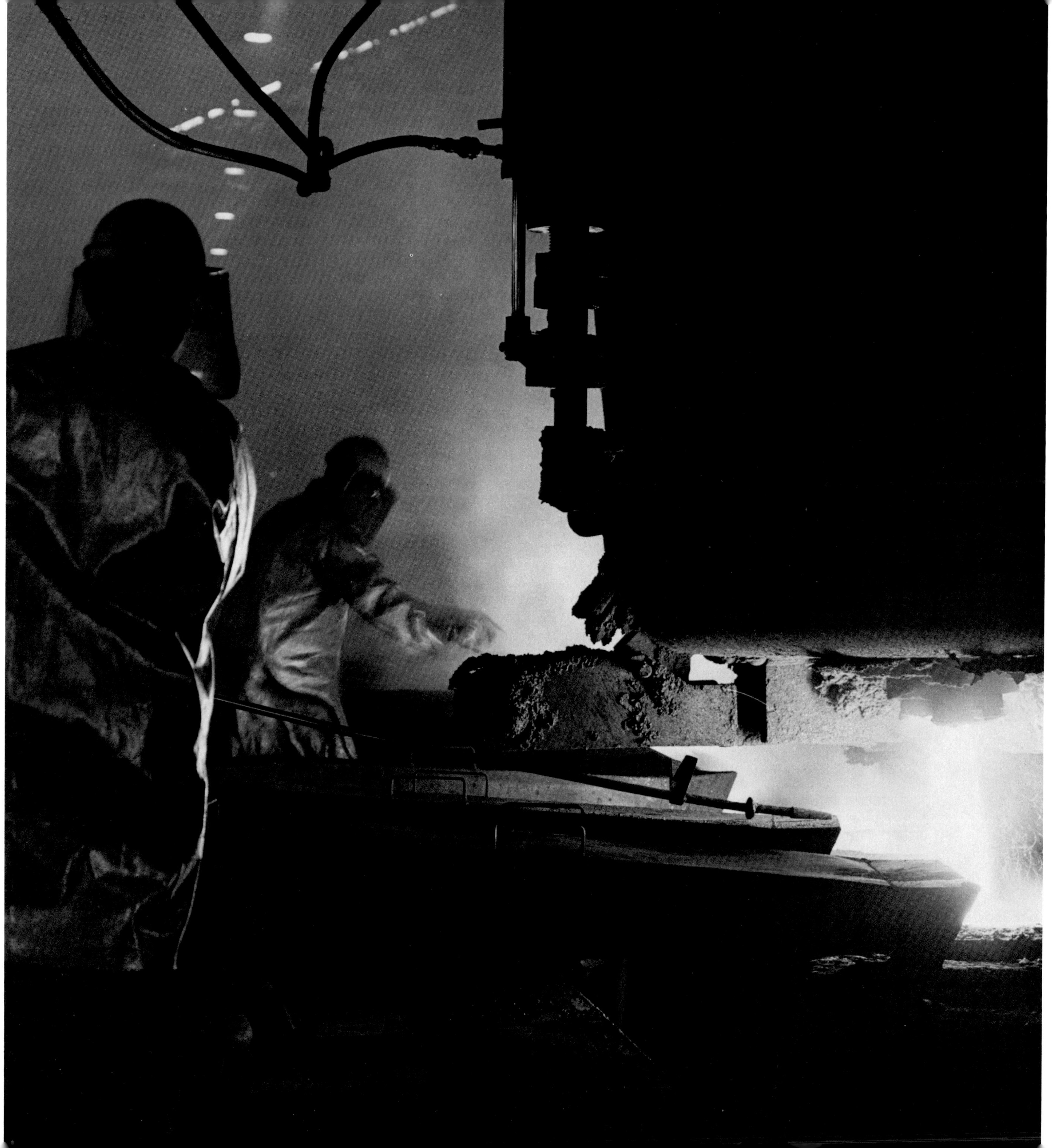

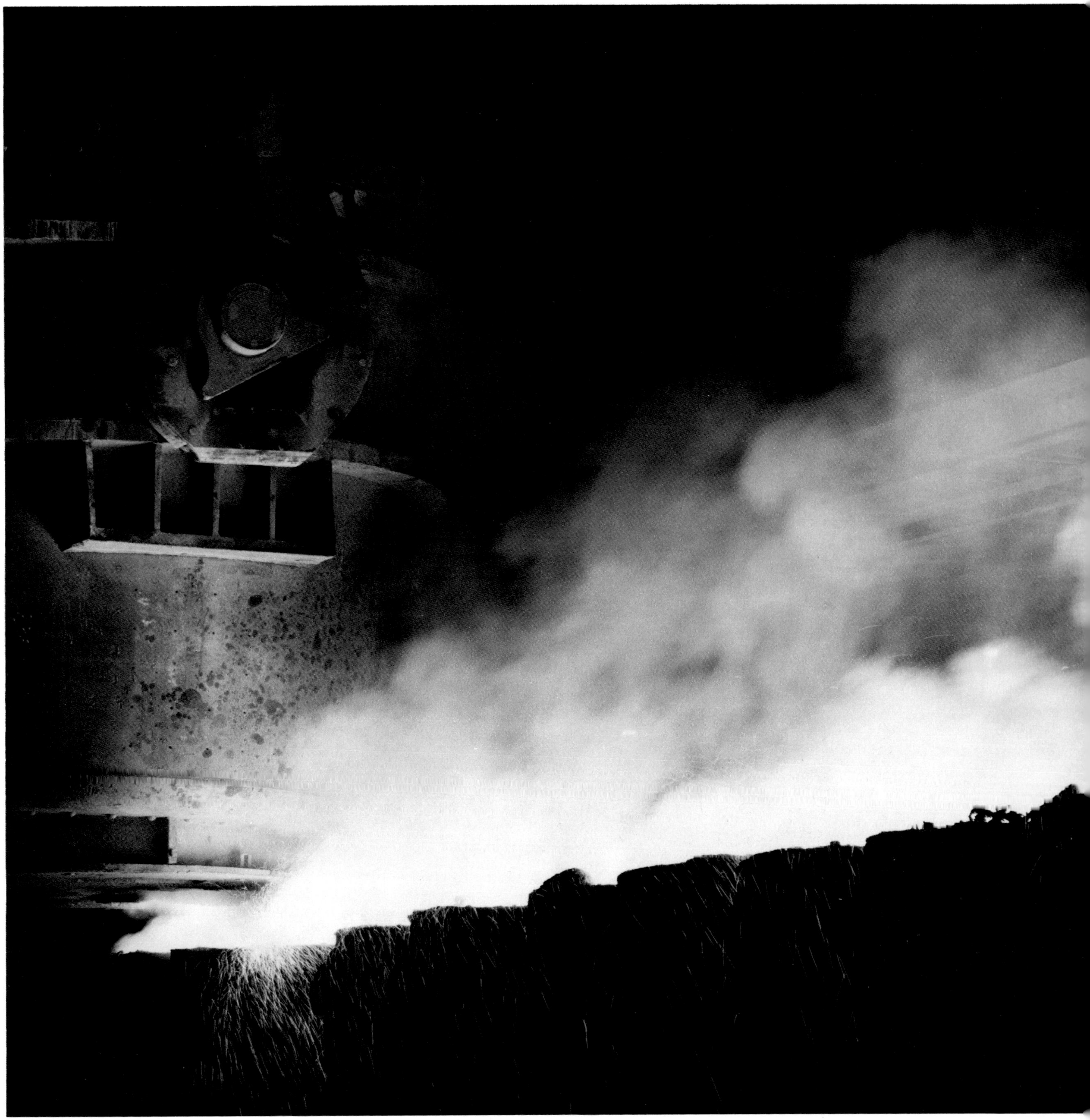

Teeming ingots.

Slagging ingots.

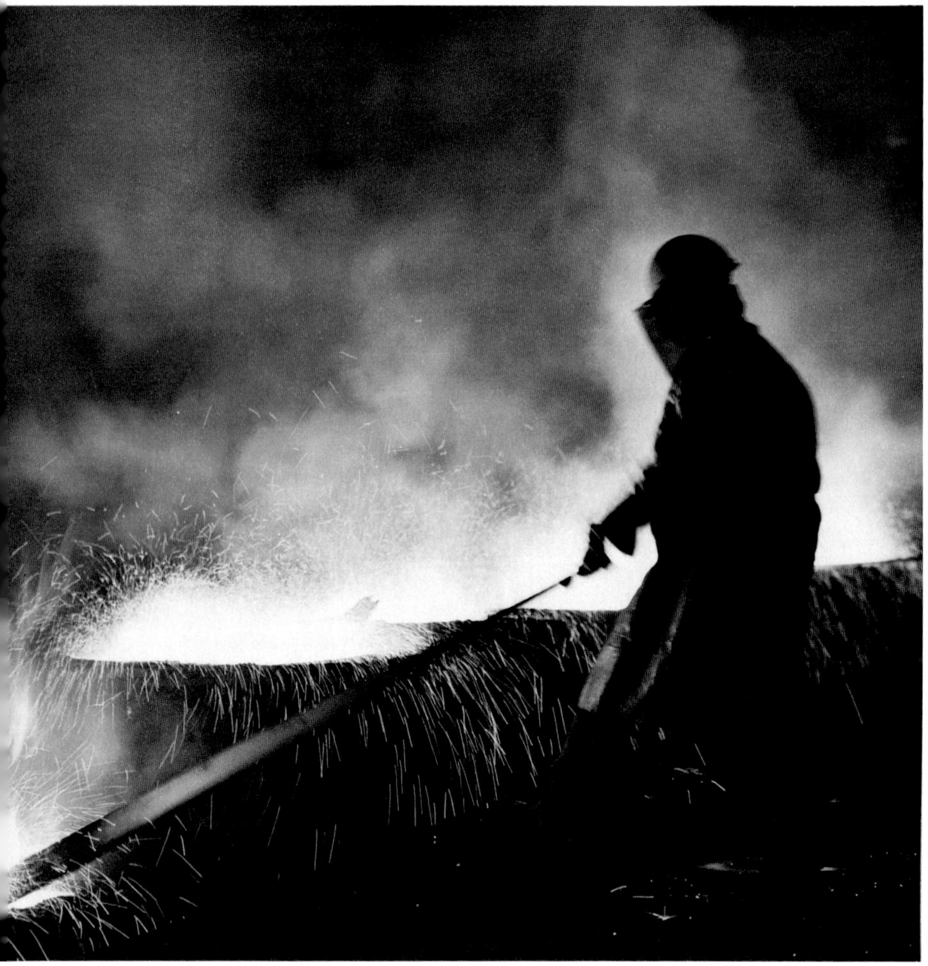

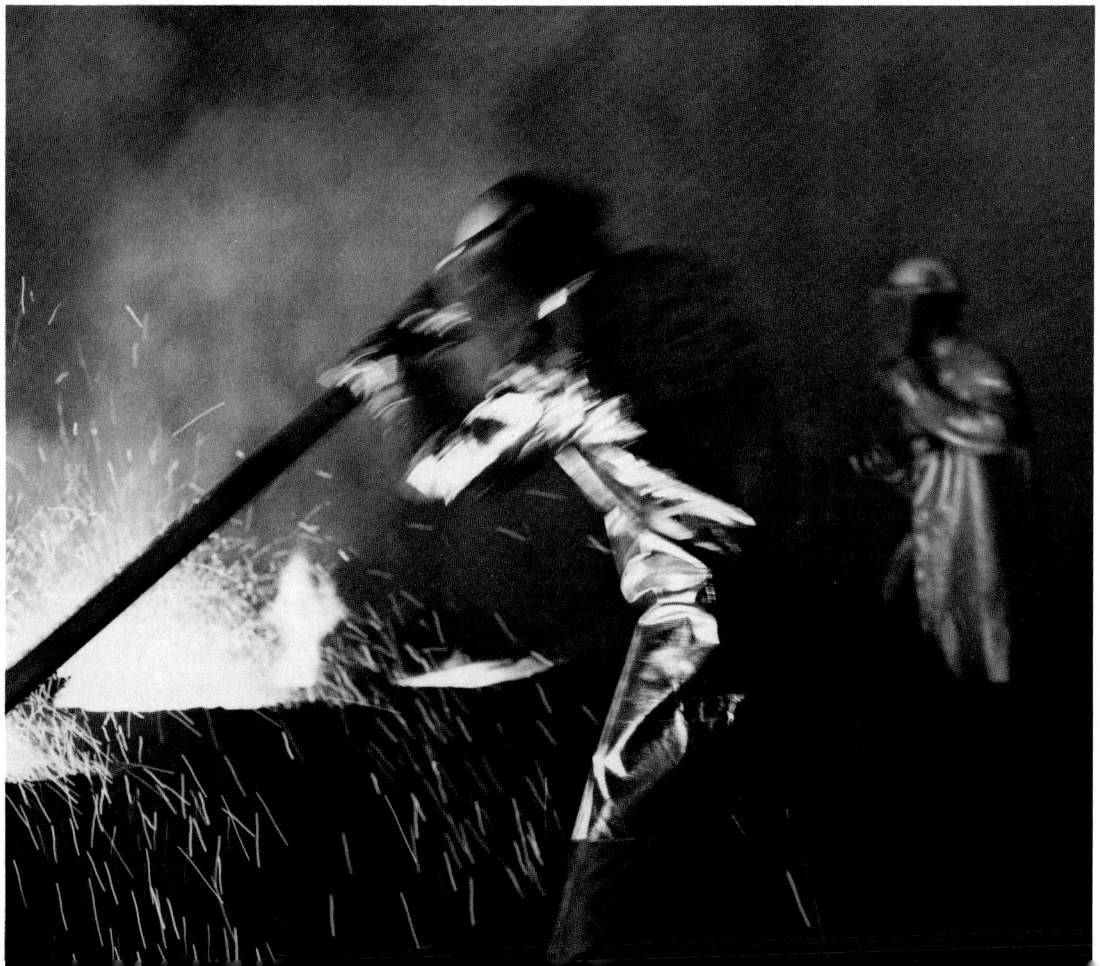

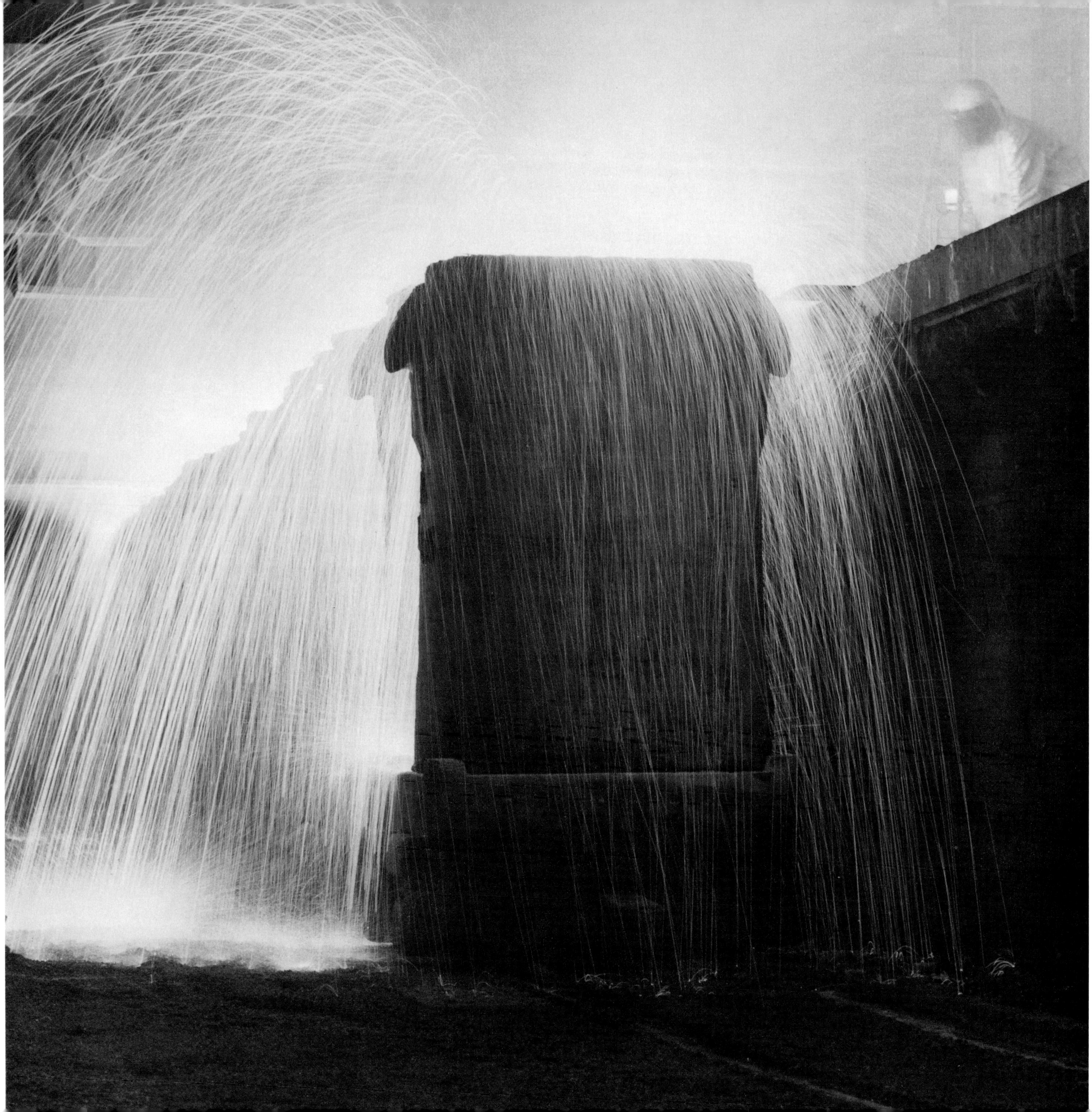

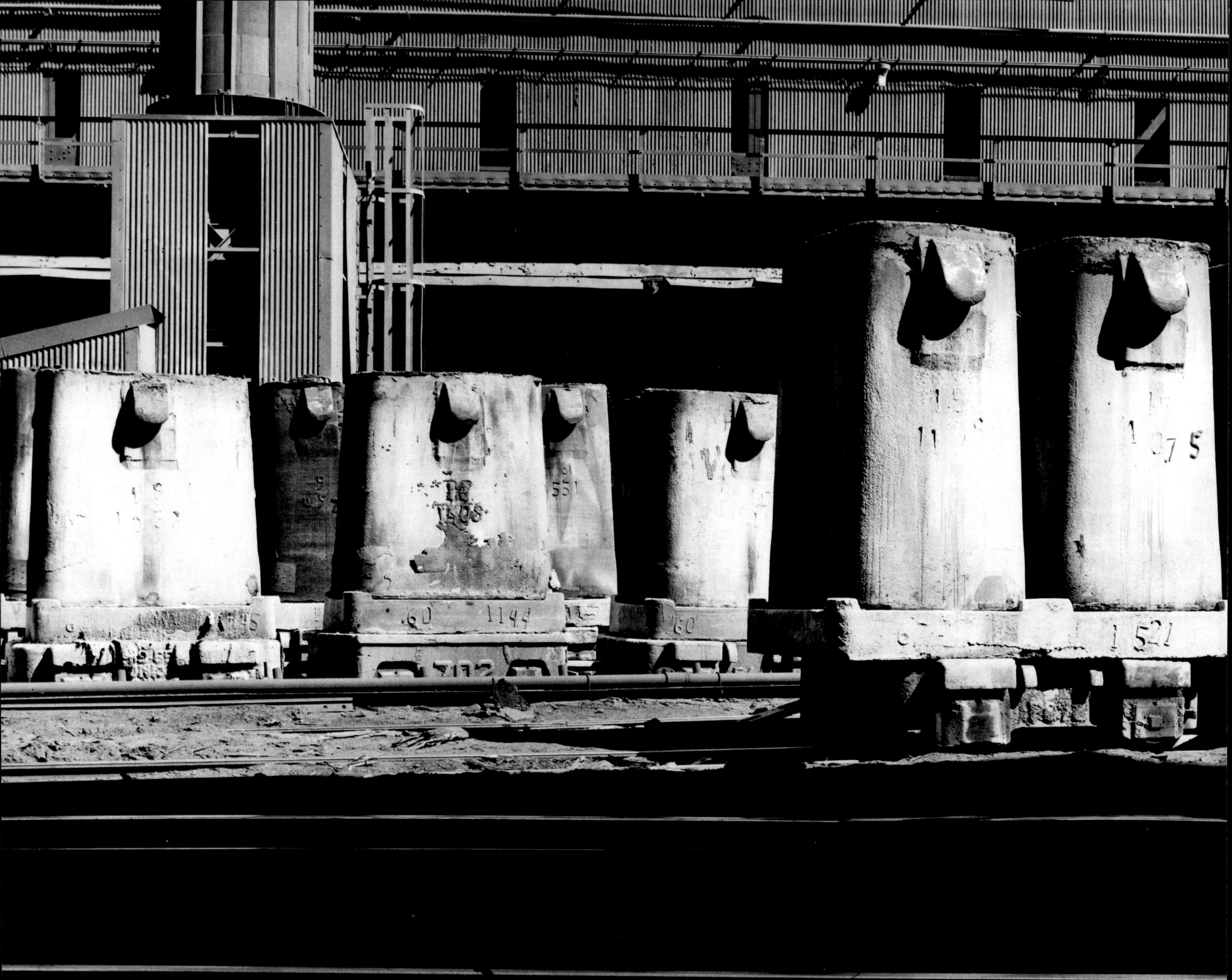
60 1144

Ingot teeming from a B.O.F.

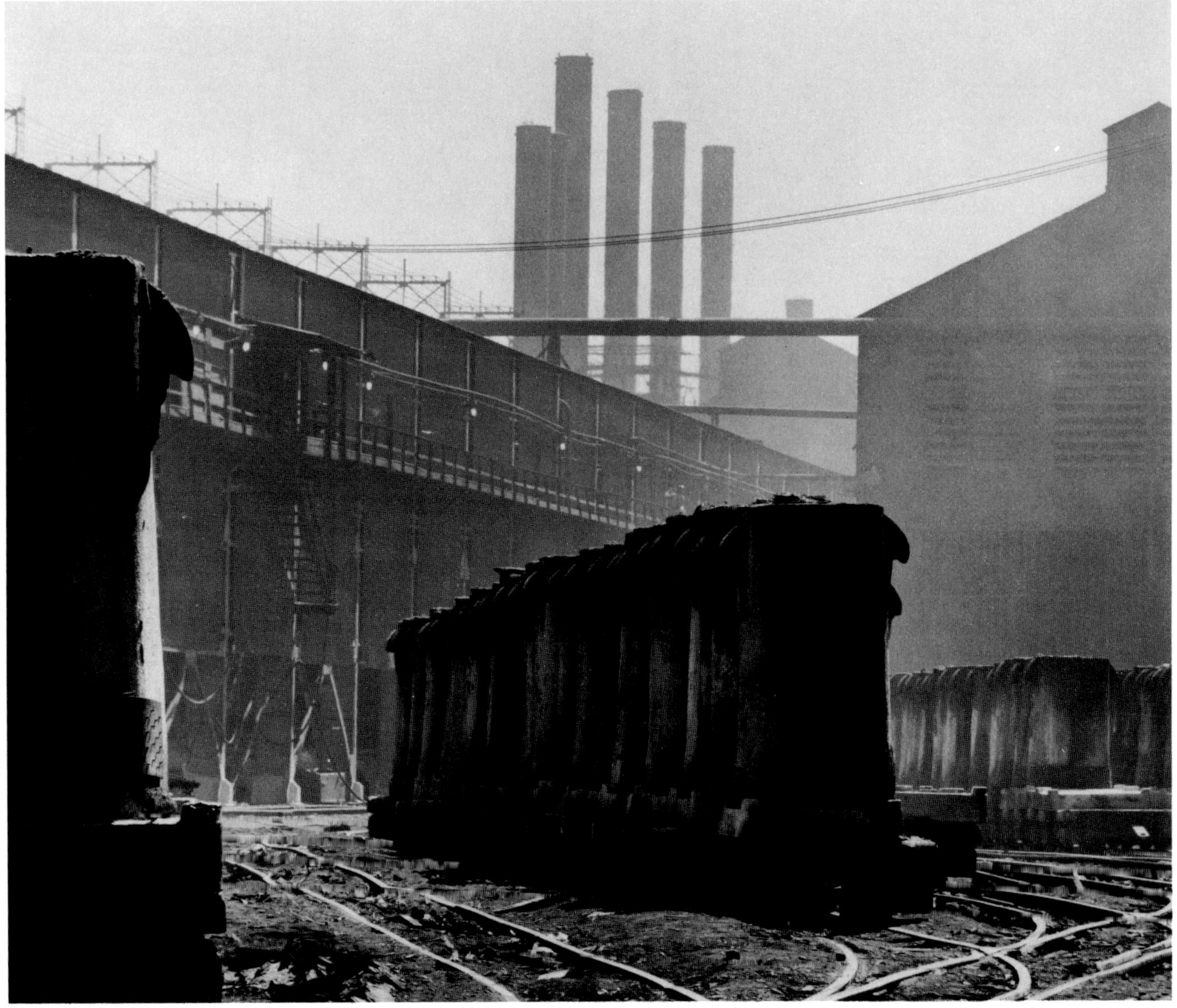

Ingot molds.

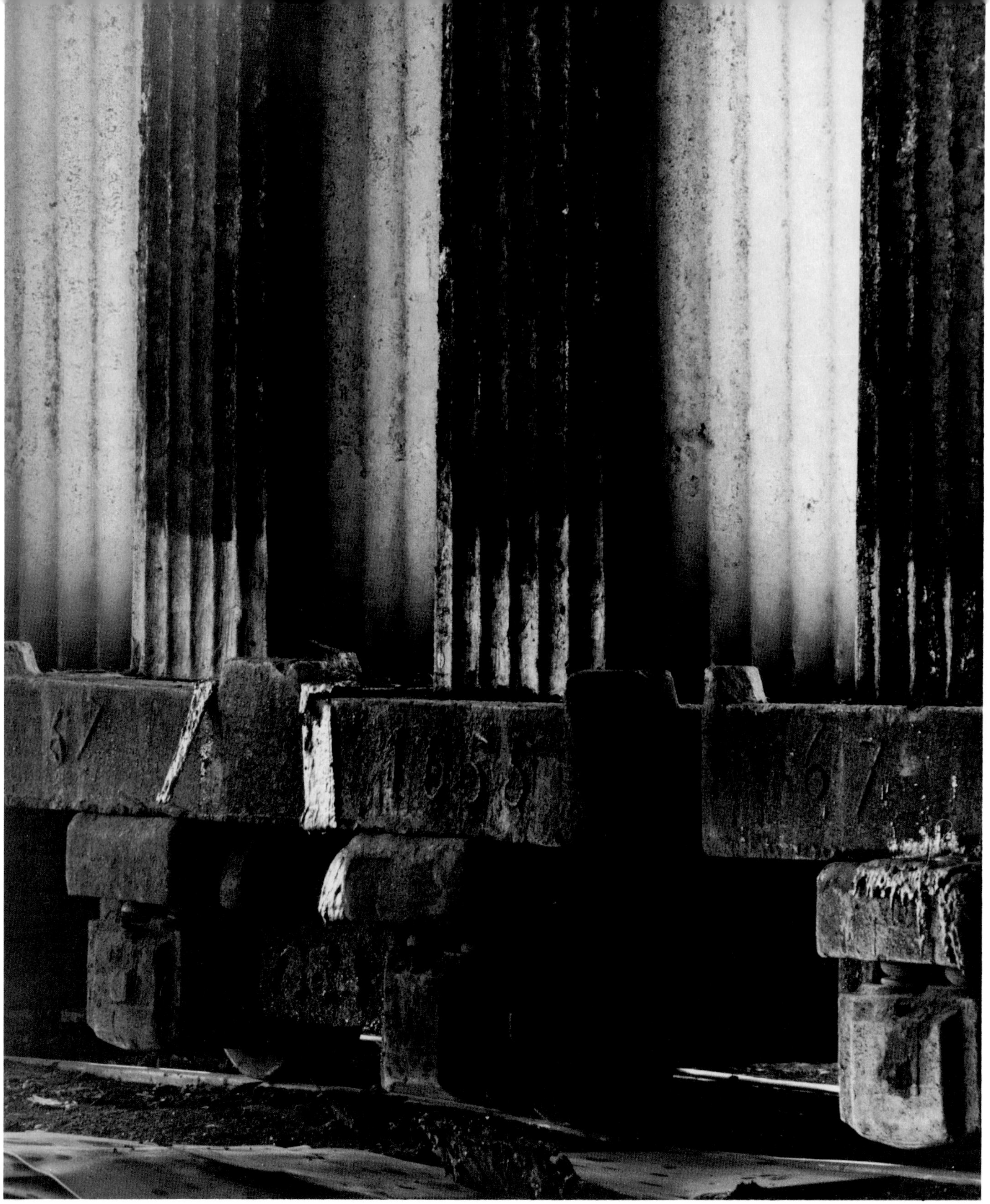

Stripped ingots.

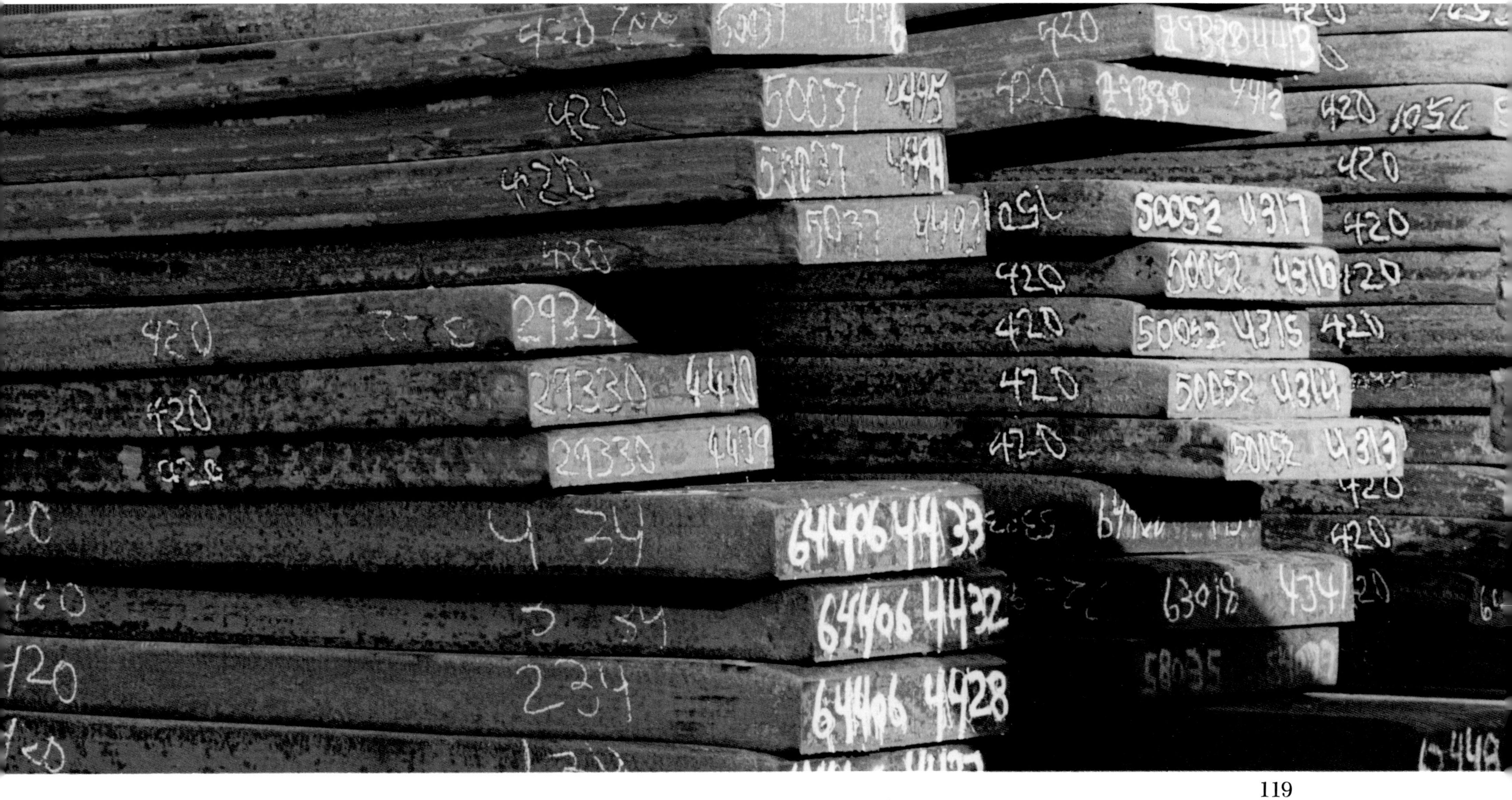
50037 4495
29330 4410
29330 4439
50052 4317
50052 4315
64406 4433
64406 4432
64406 4428
63018 4341
420

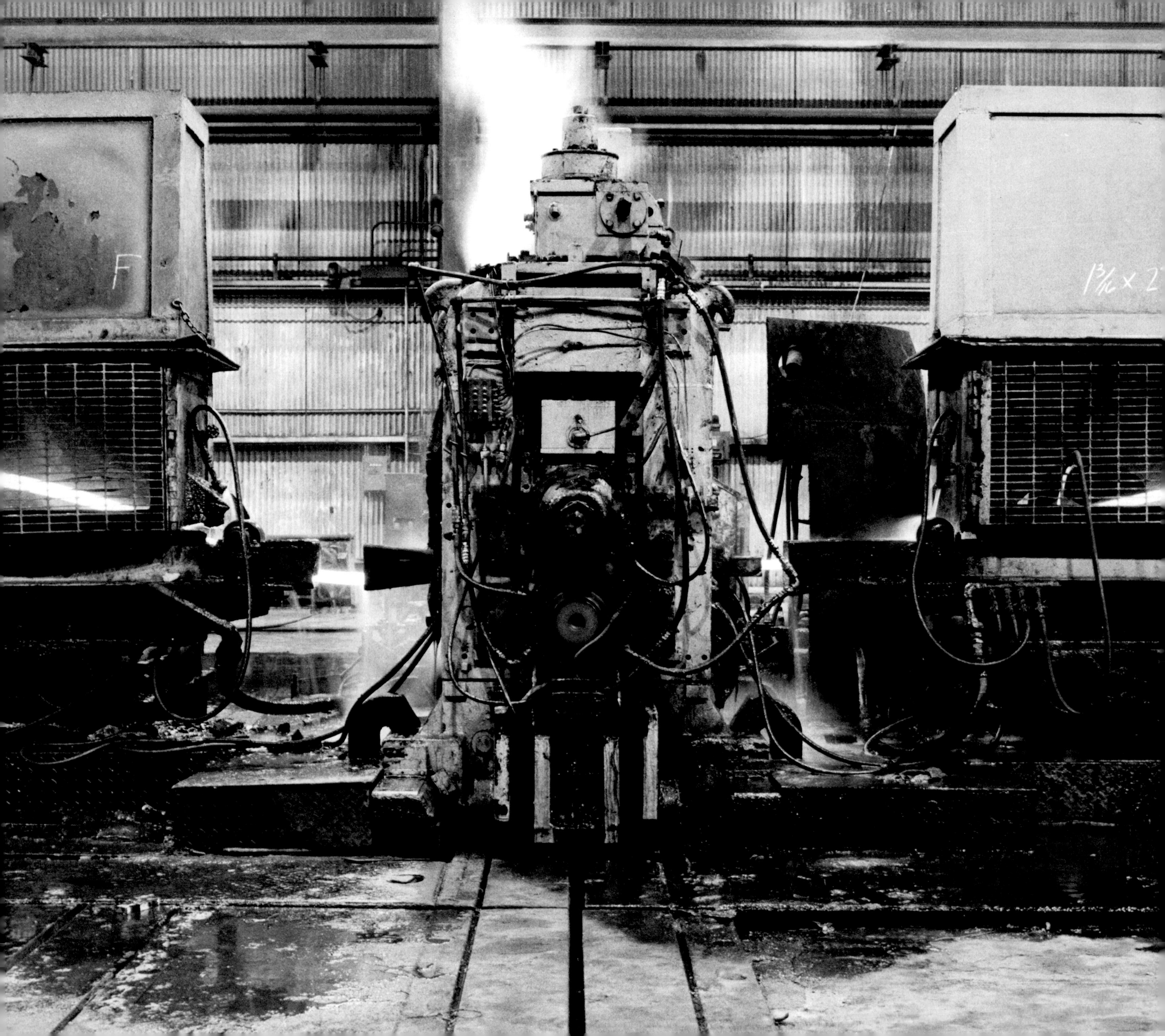
F
13/16 × 2

12-inch bar mill.

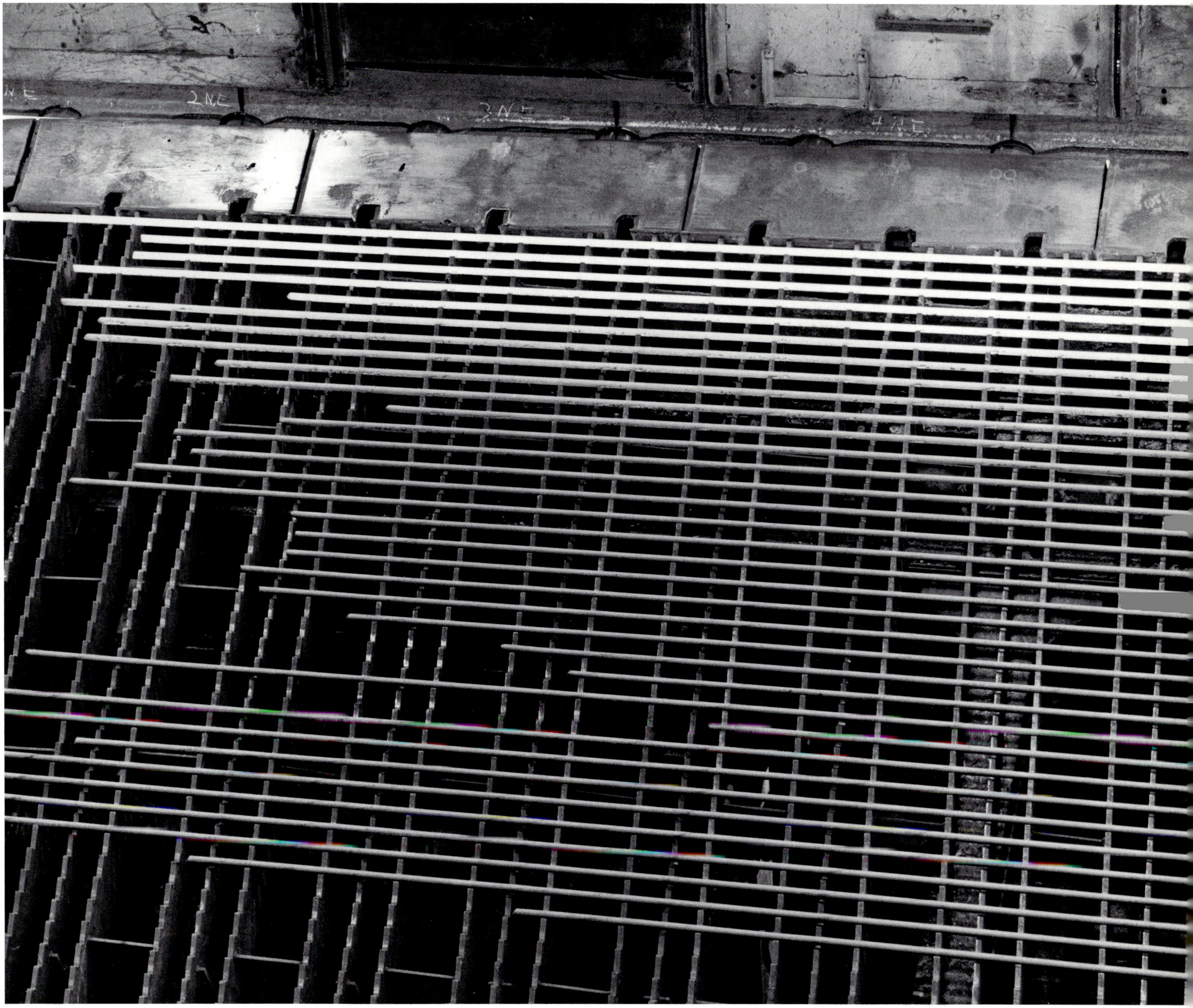

Hot table.

lling stand.

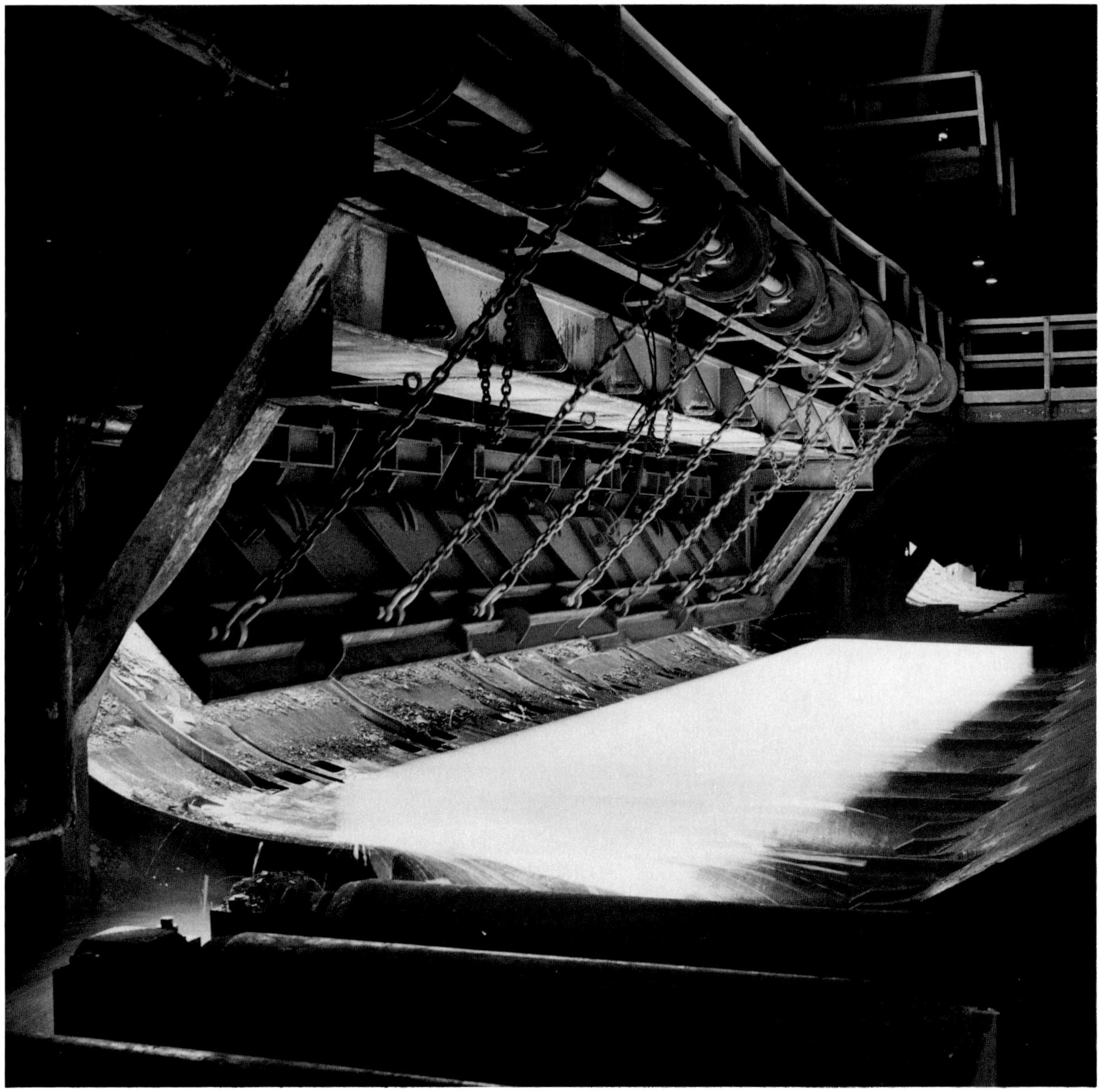

Slab expelled onto a conveyor table from a reheating furnace.

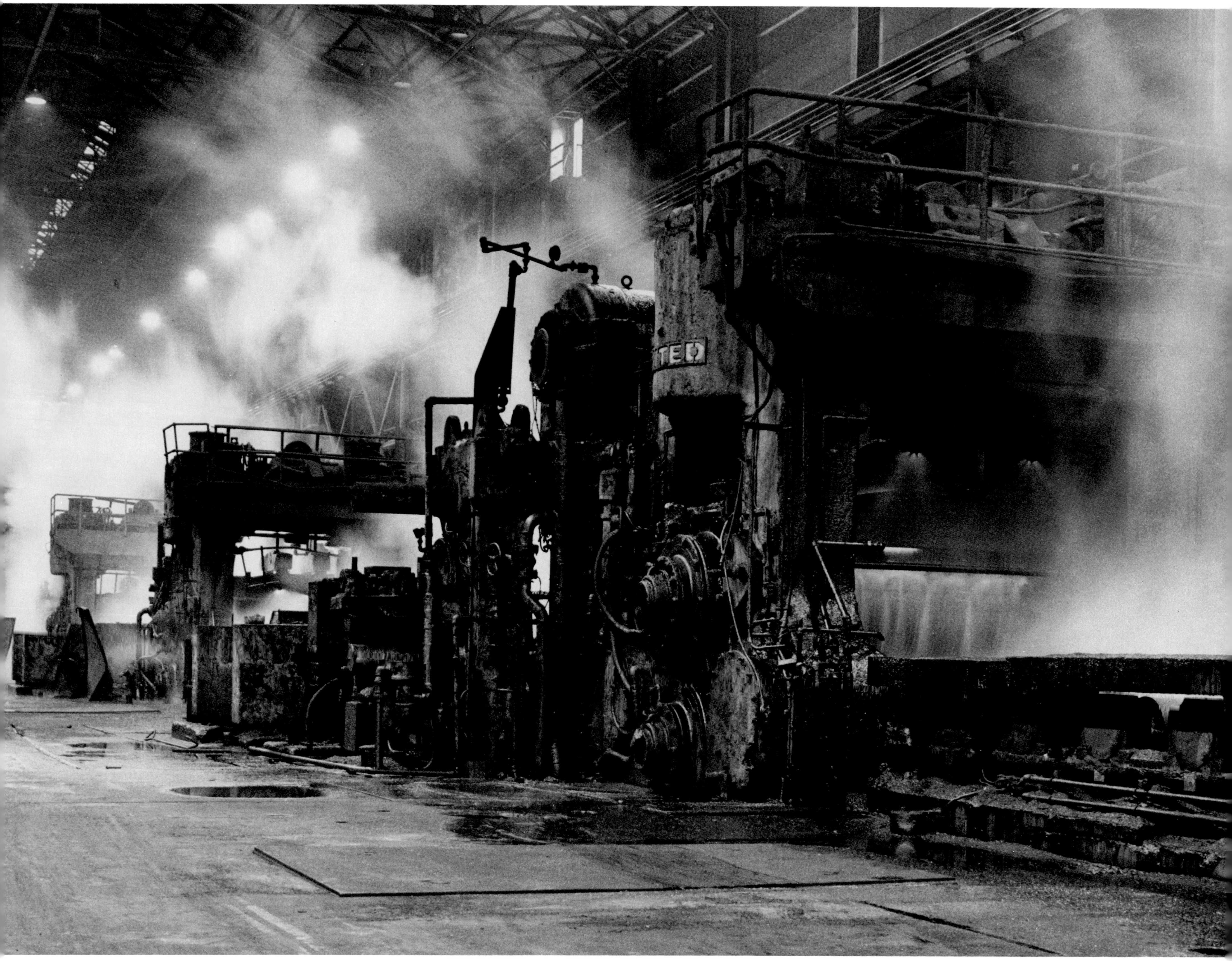

Overleaf: 2 high horizontal roughing stands; slab exiting a rougher on an 80-inch hot-strip mill.

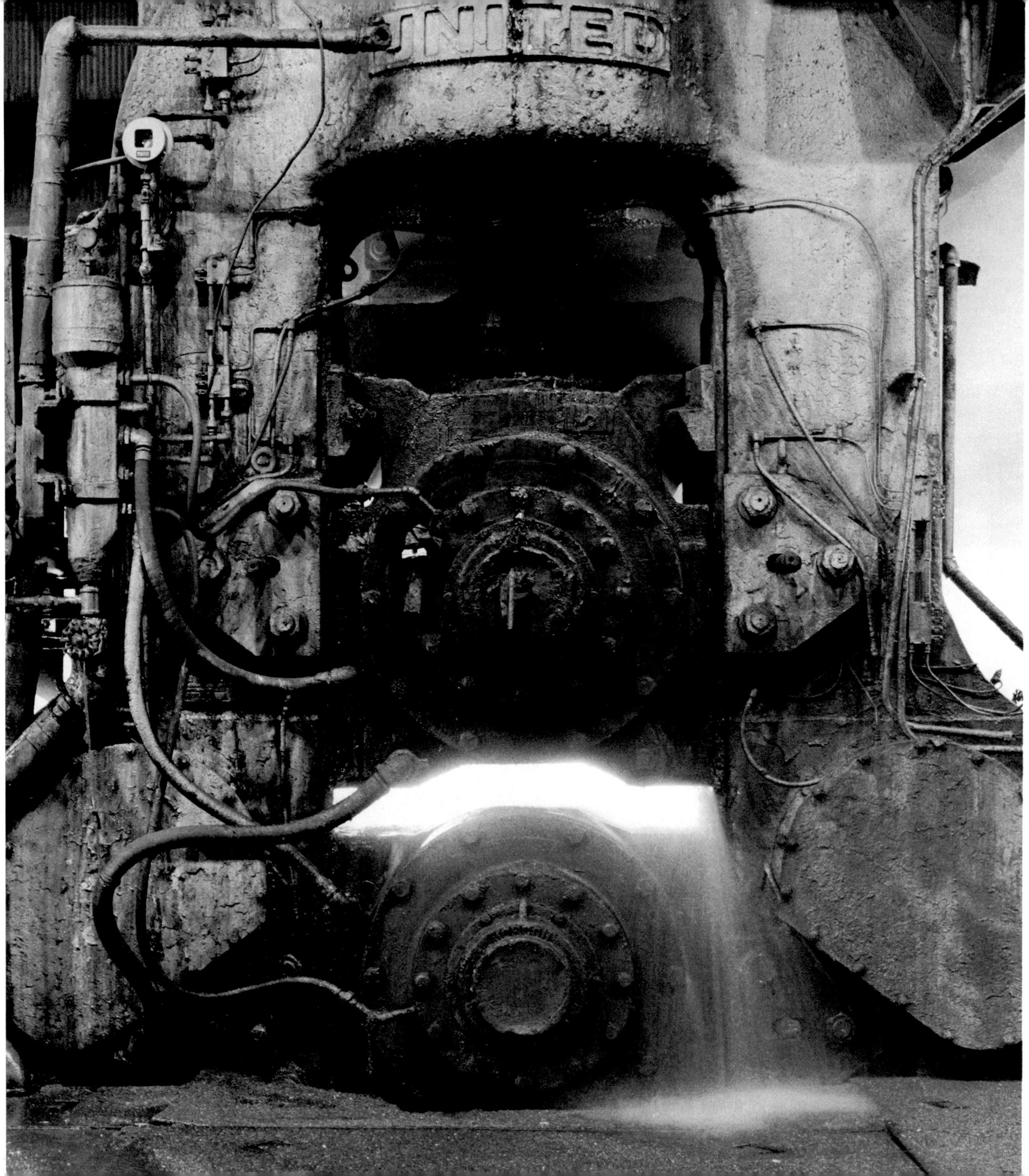
UNITED

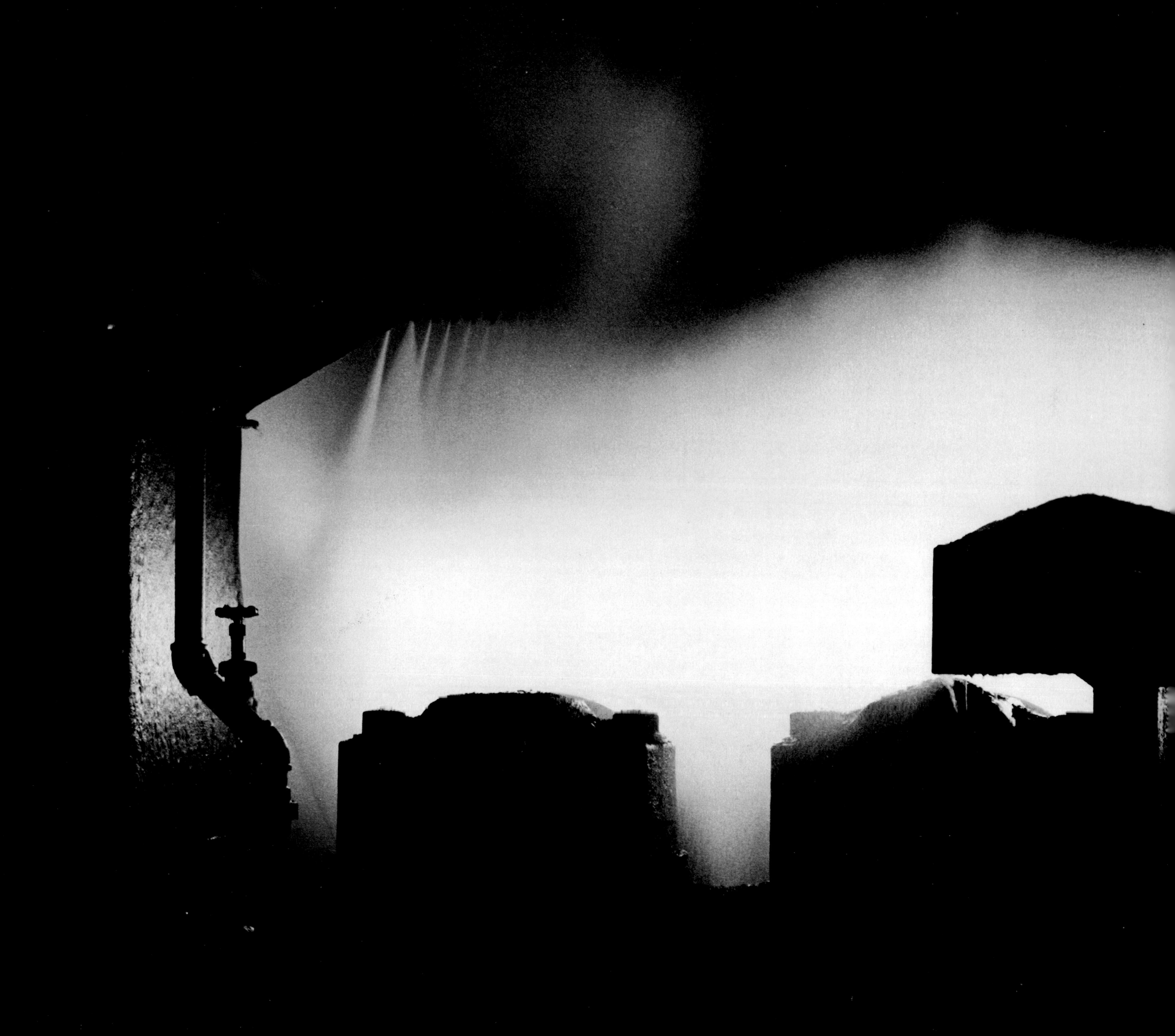

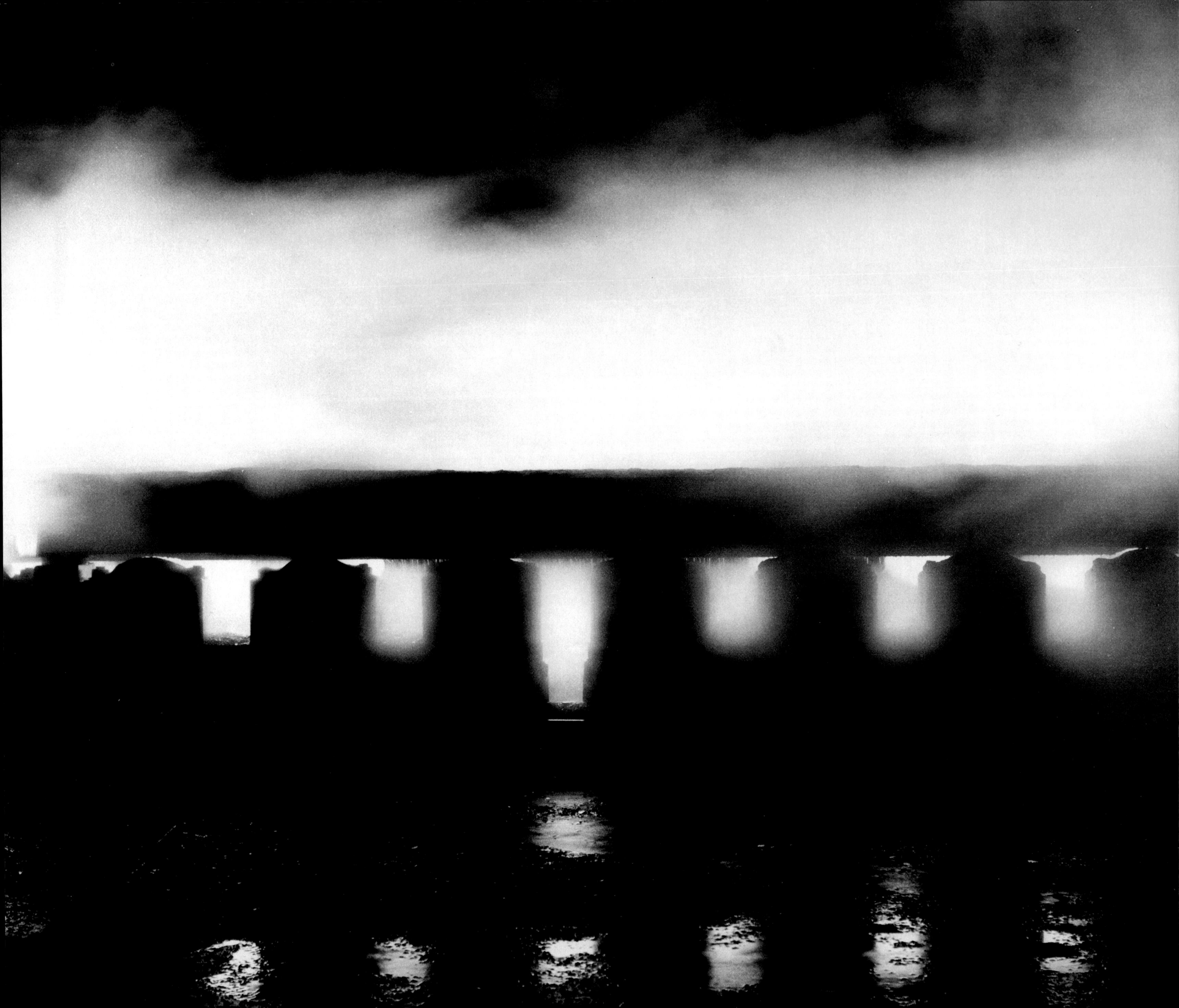

Preceding pages: conveyor tables in an 80-inch hot-strip mill.

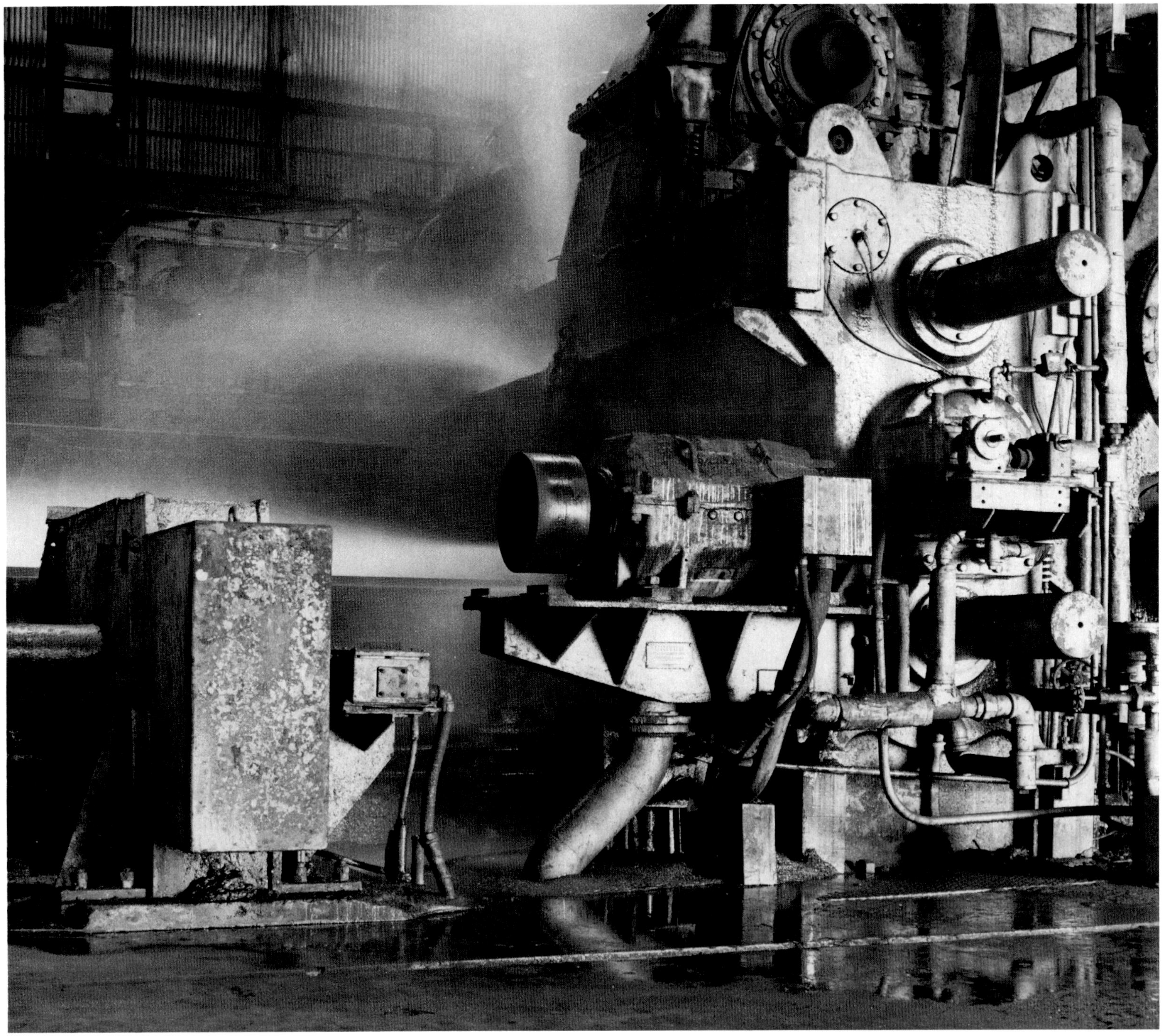

Details of a rolling stand in an 80-inch hot-strip mill.

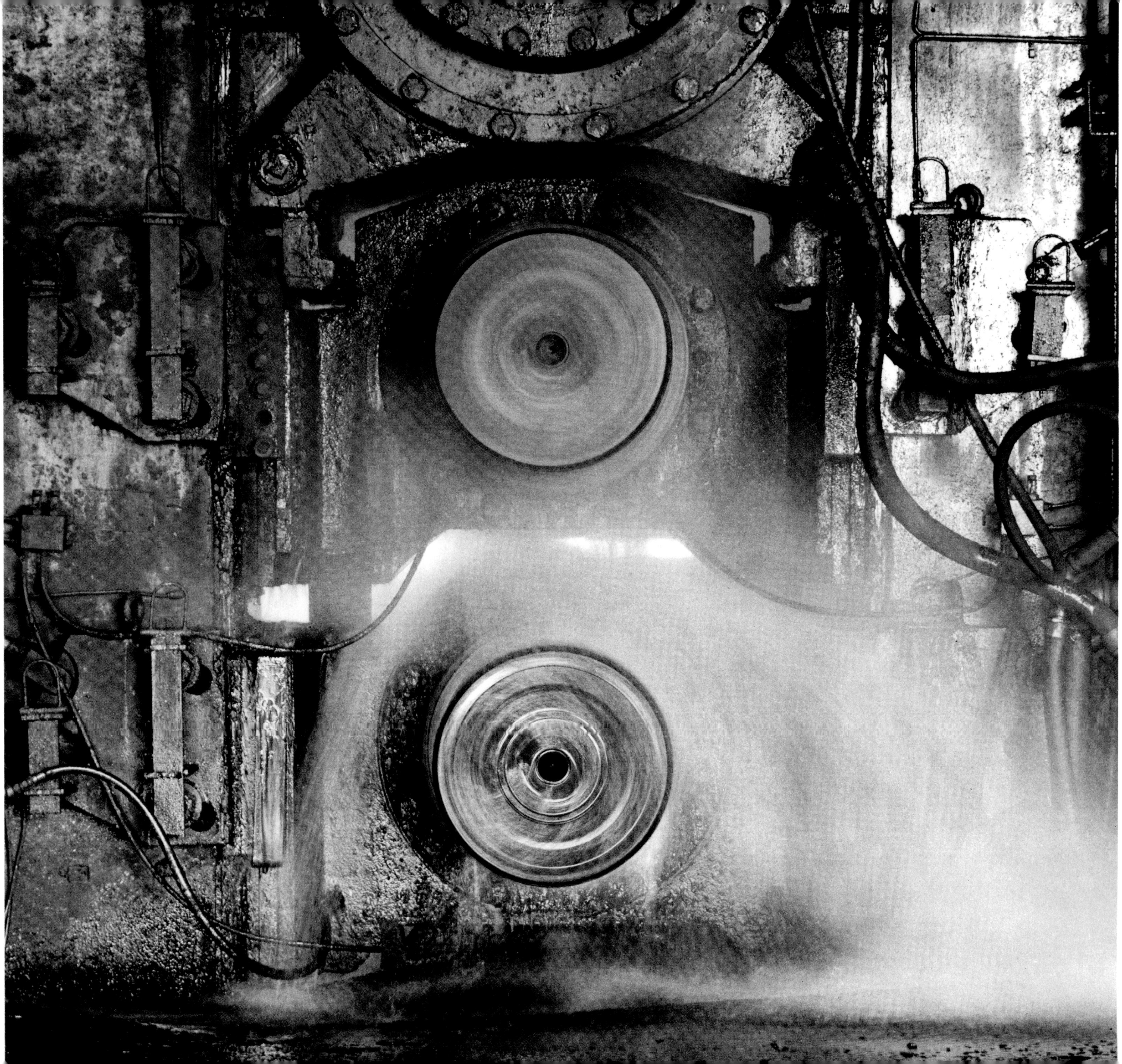

80-inch hot-strip mill.

Conveyor table.

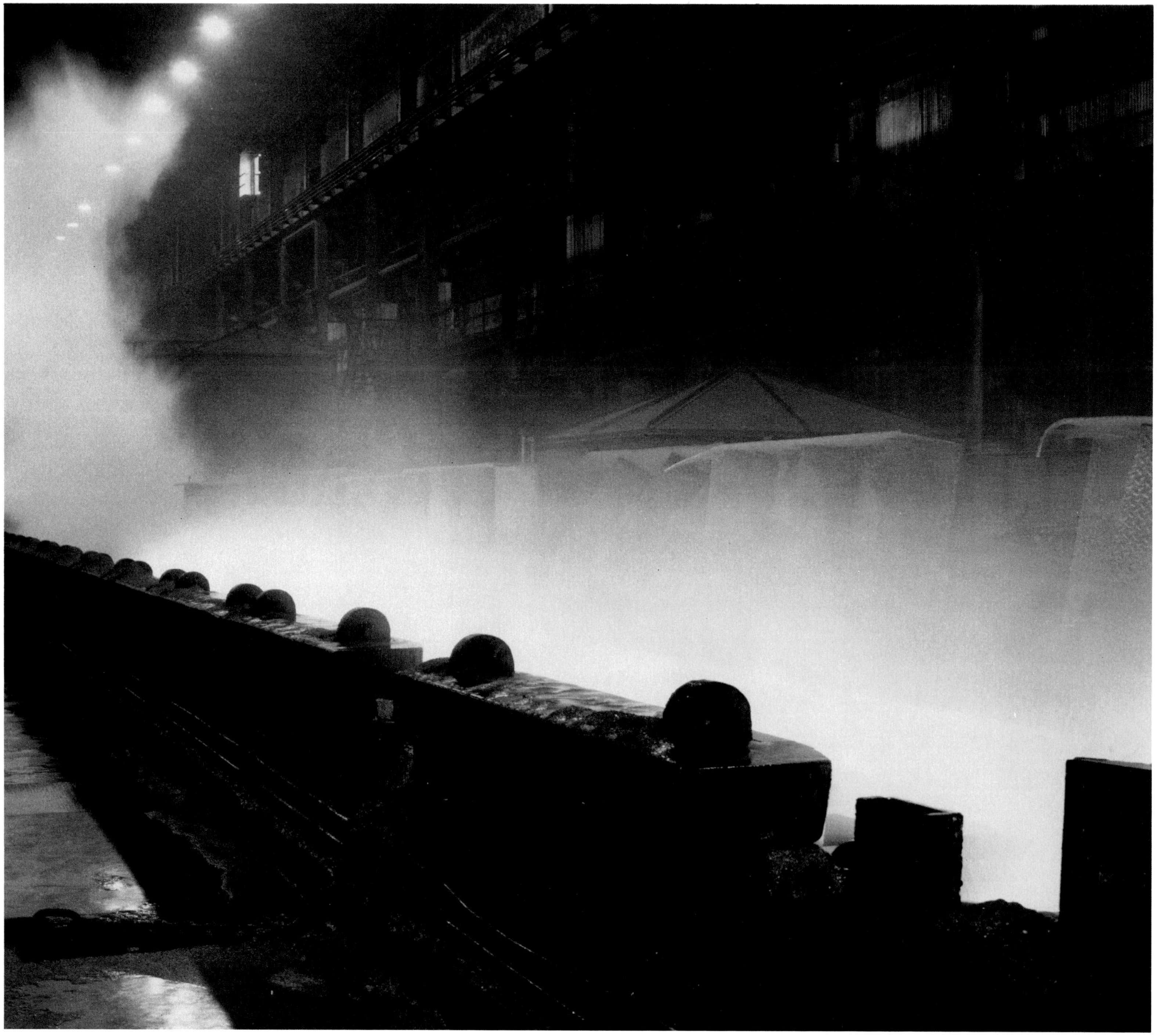

Finishing stands.

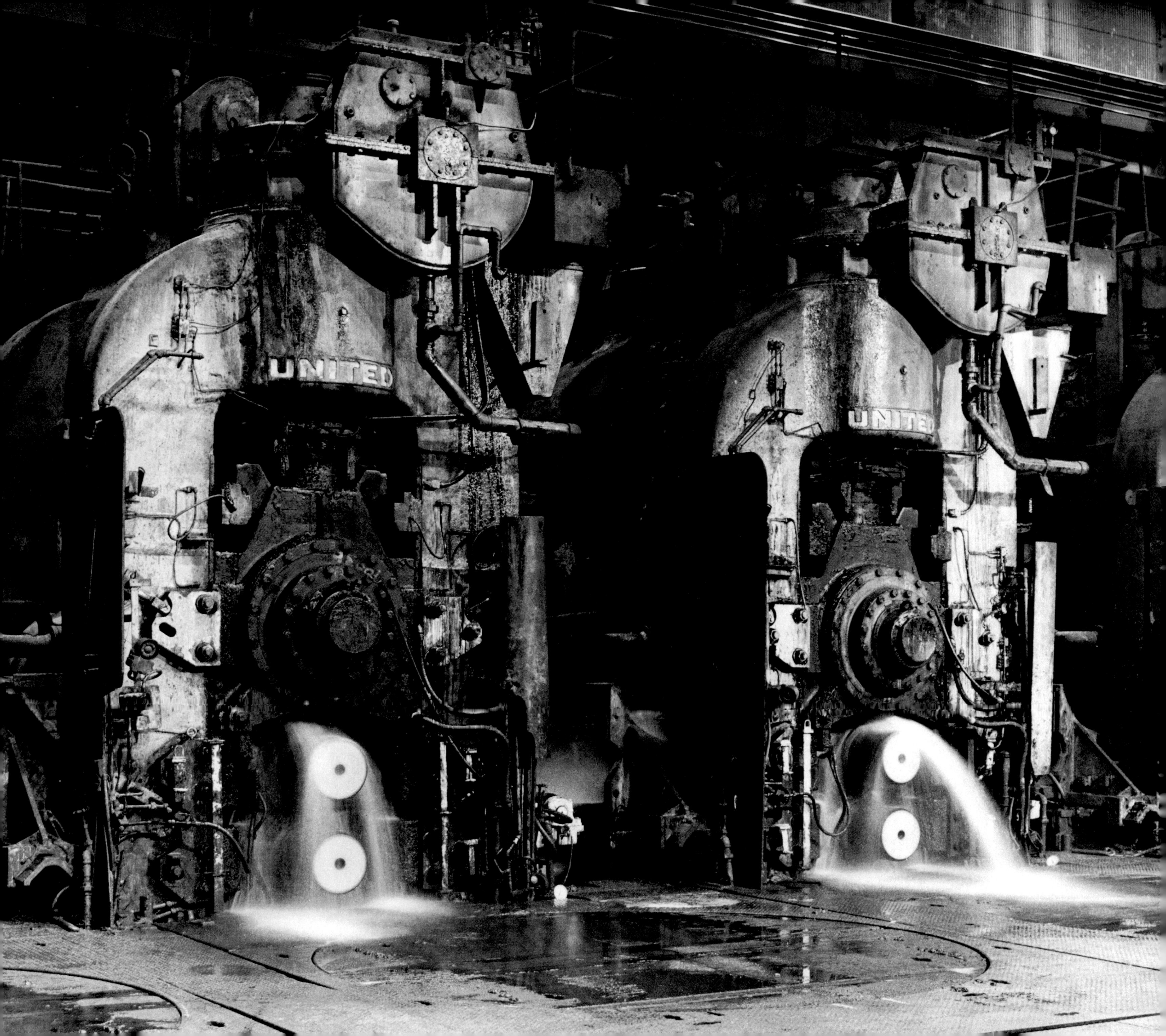
UNITED
UNITED

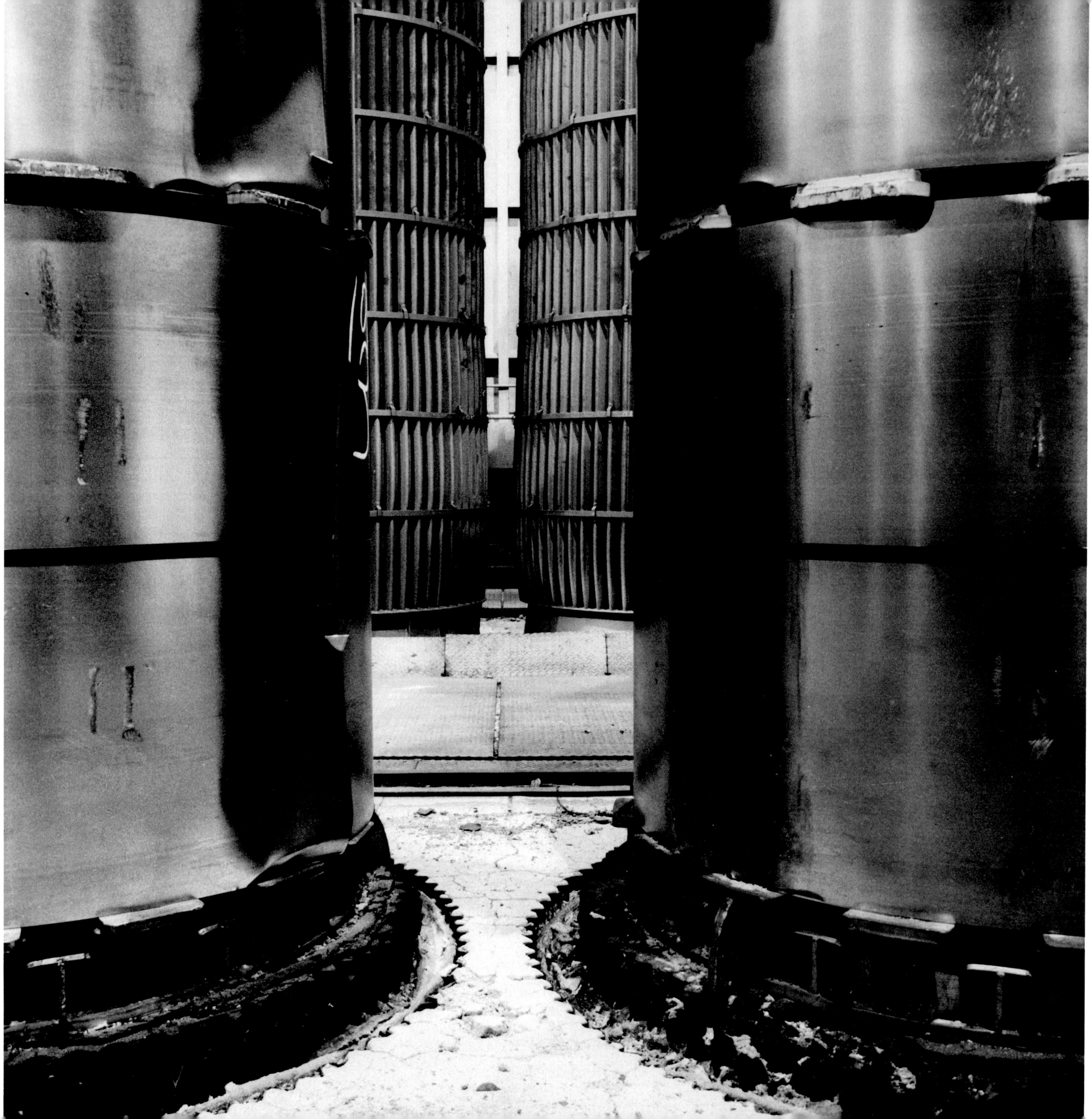

Open-coil annealing.

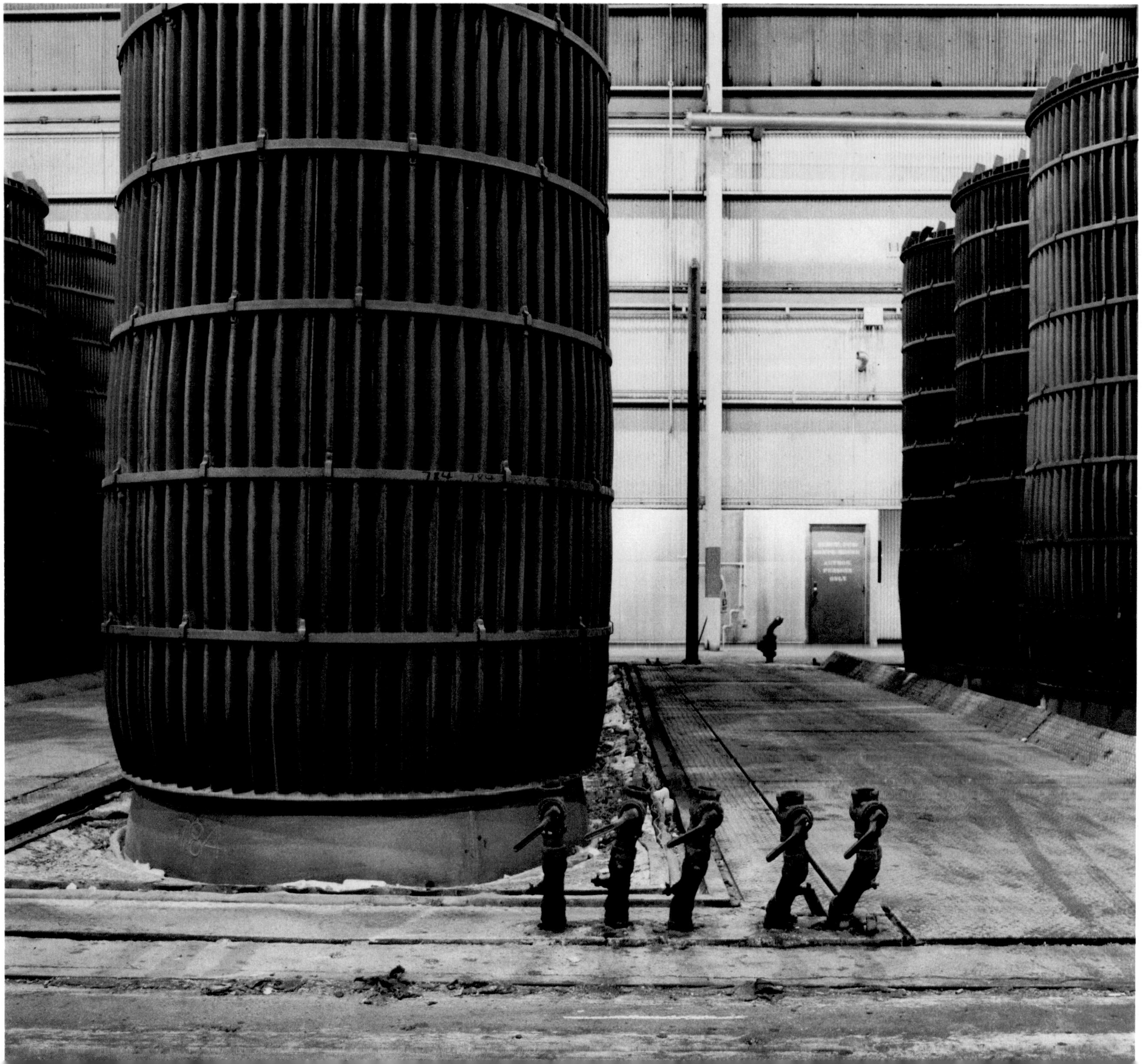

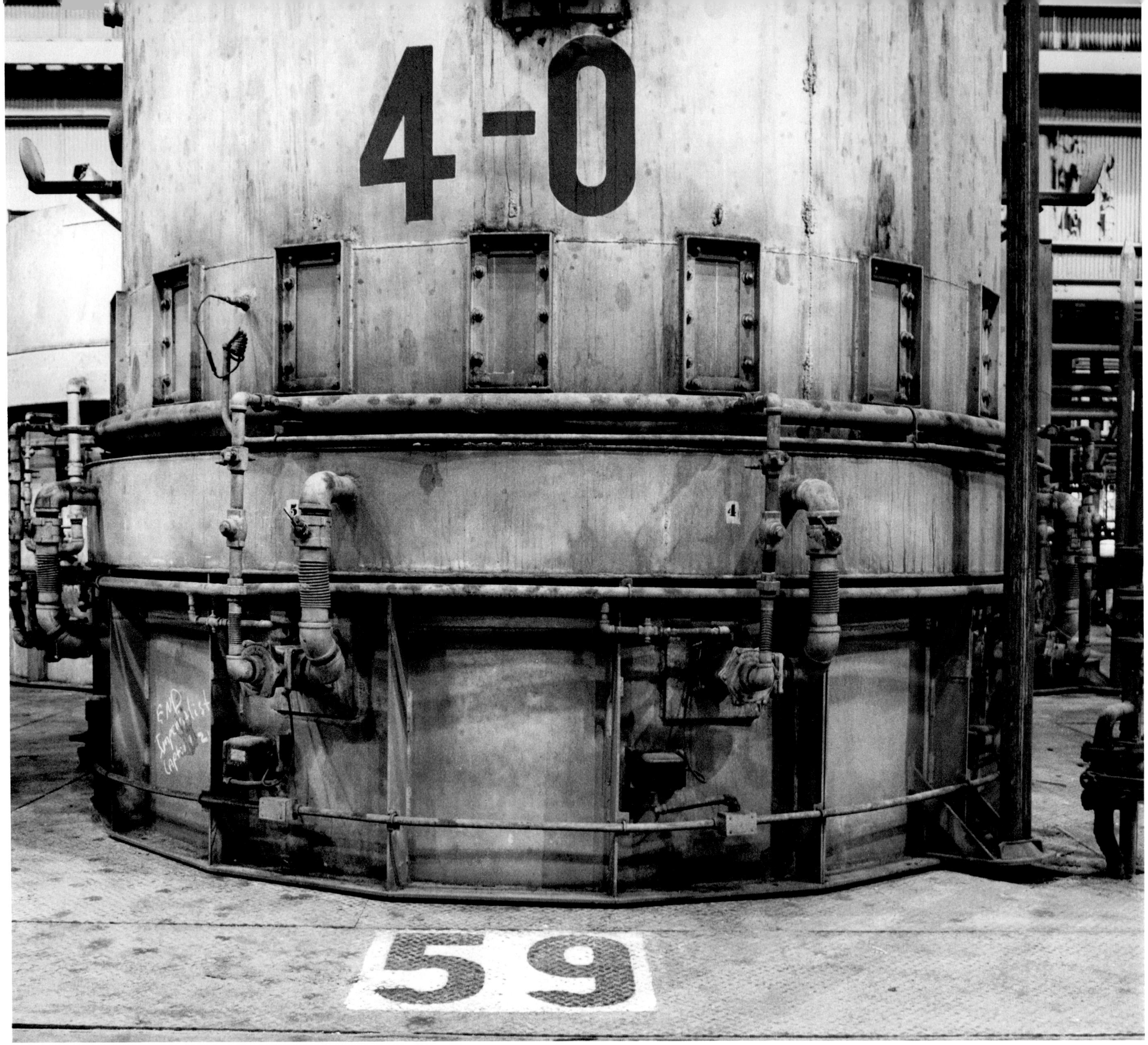

Annealing furnace.

Detail on a galvanizing lin

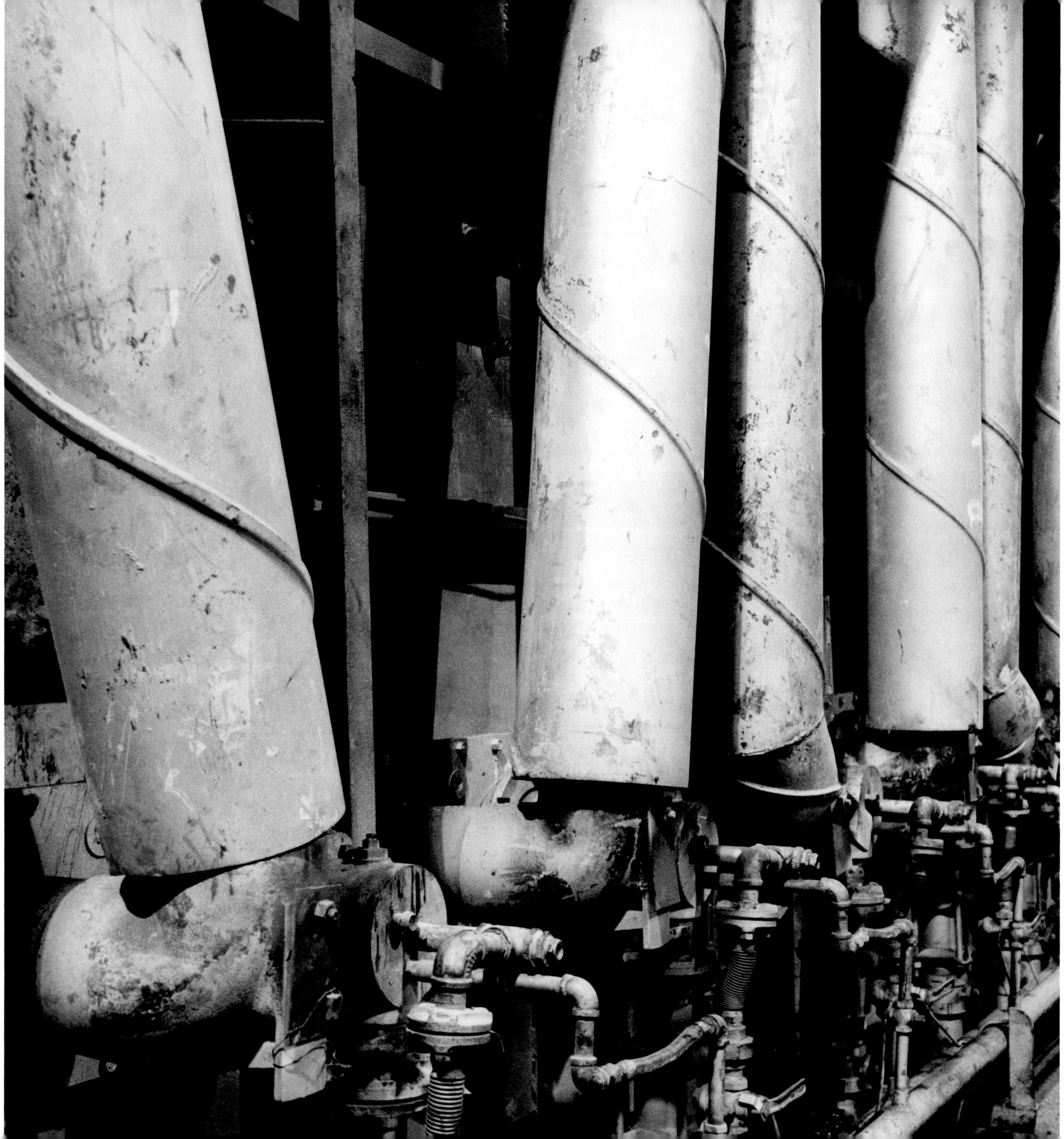

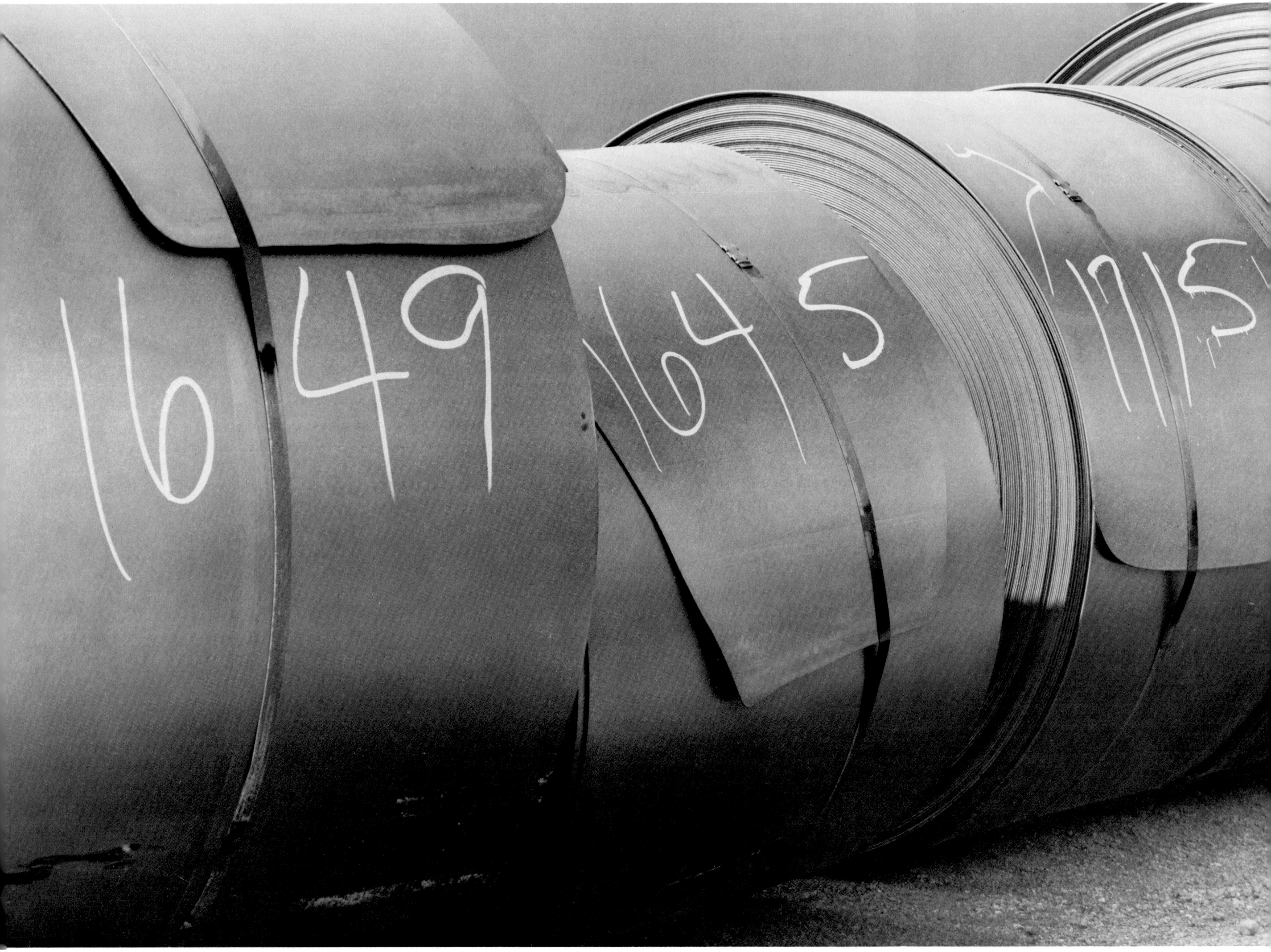
1649
1645

Coils of hot-rolled steel.

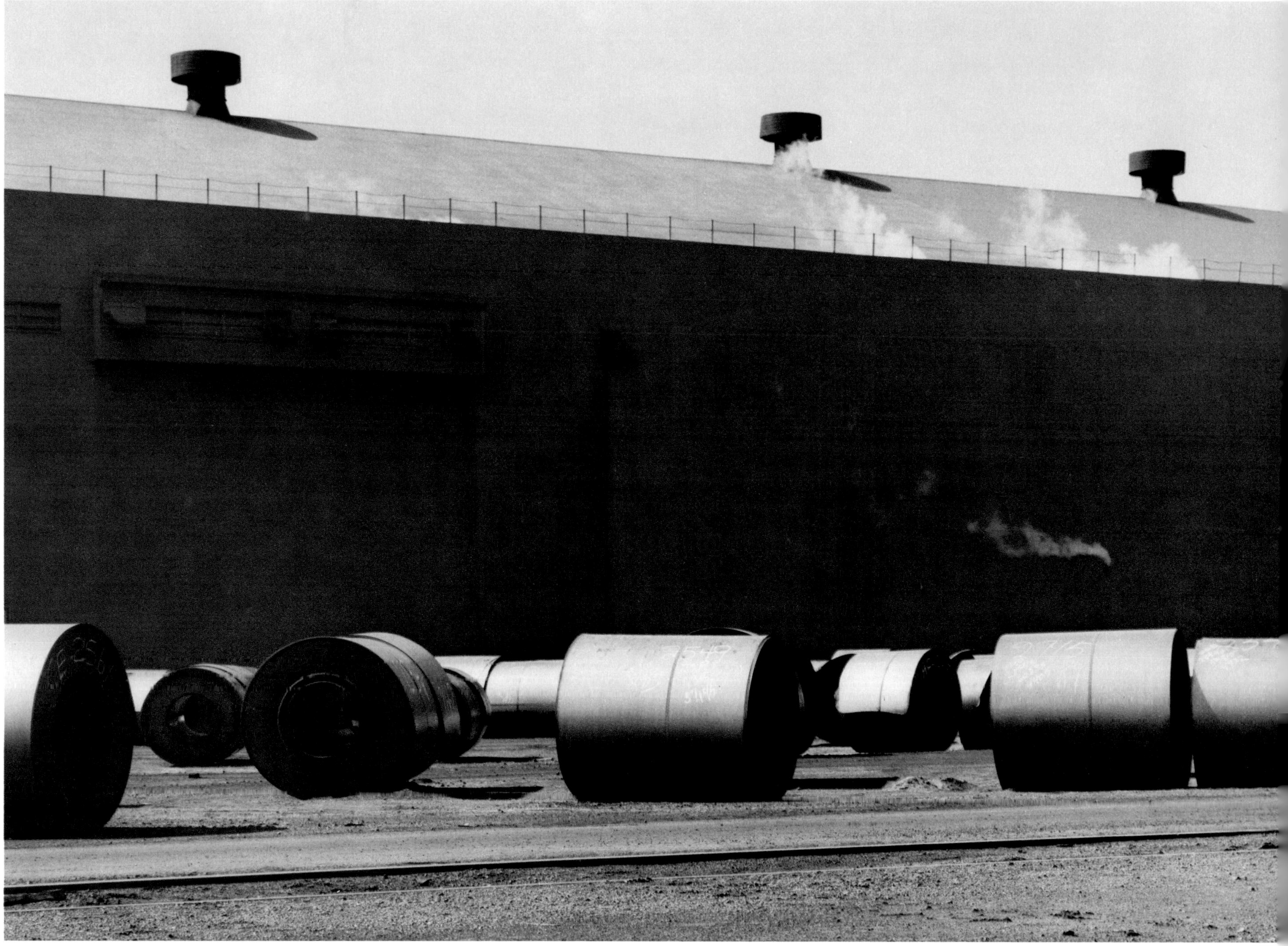

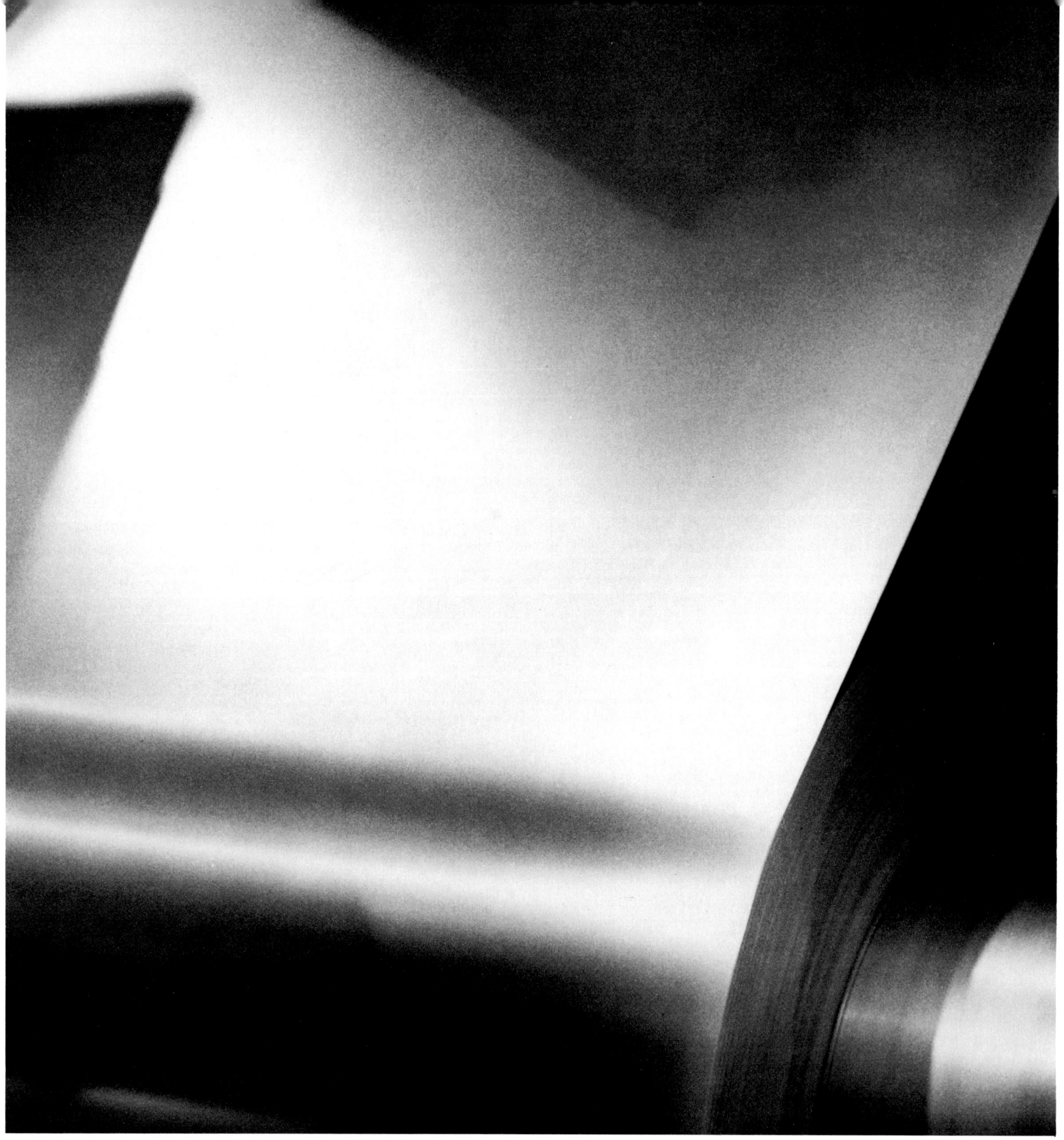

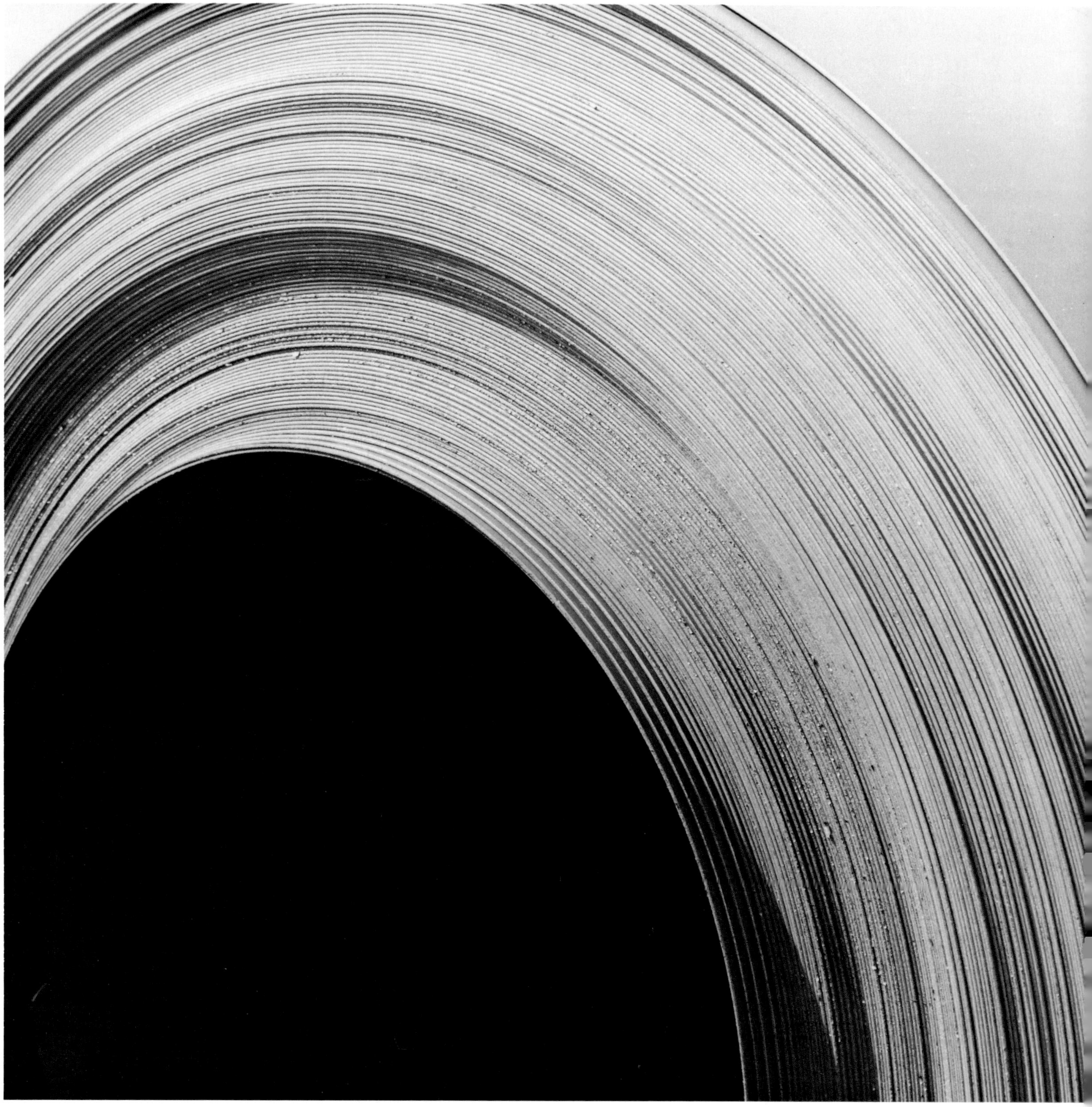

Coils of cold-rolled sheet steel.

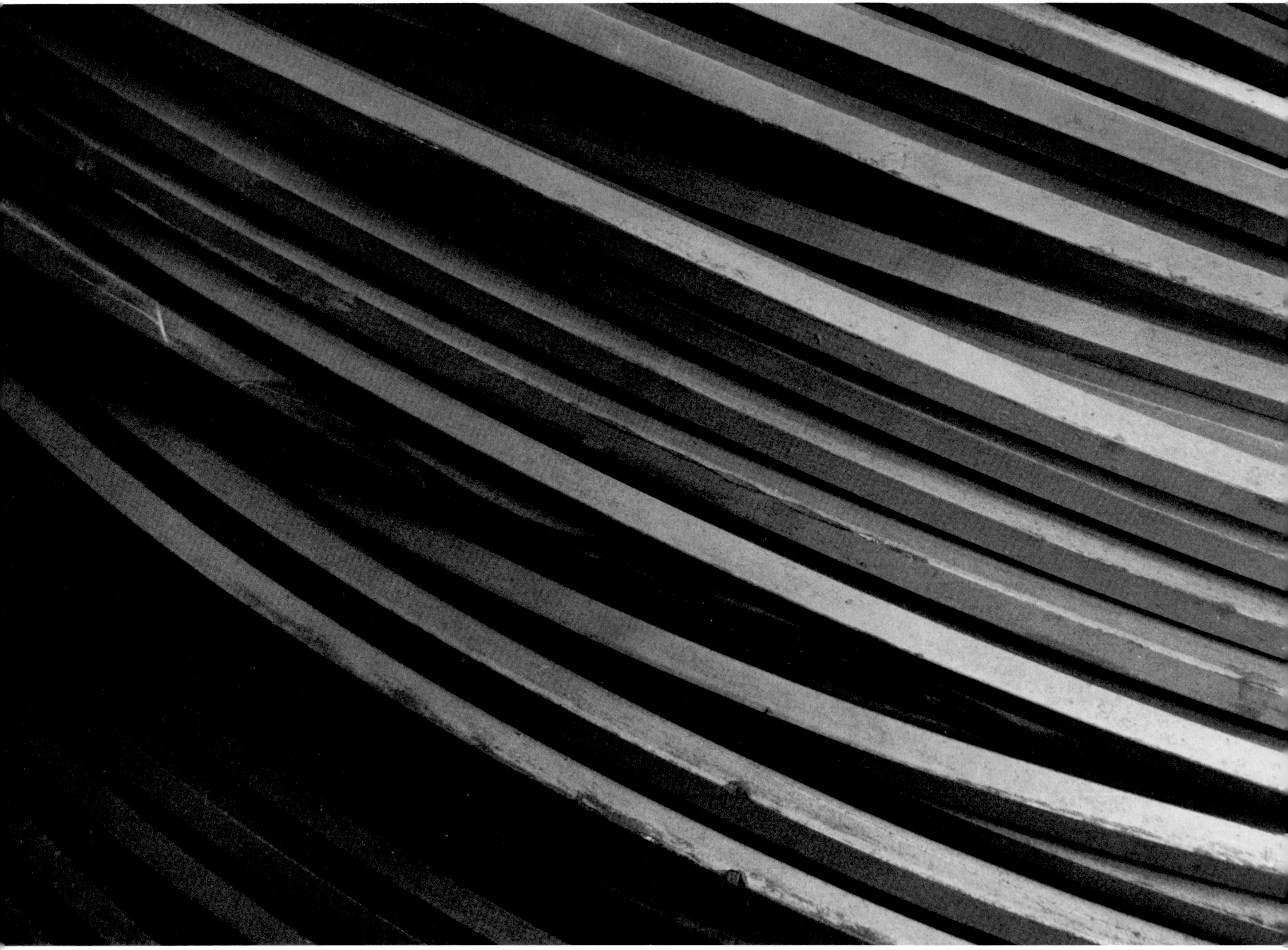

Reinforcing rods.

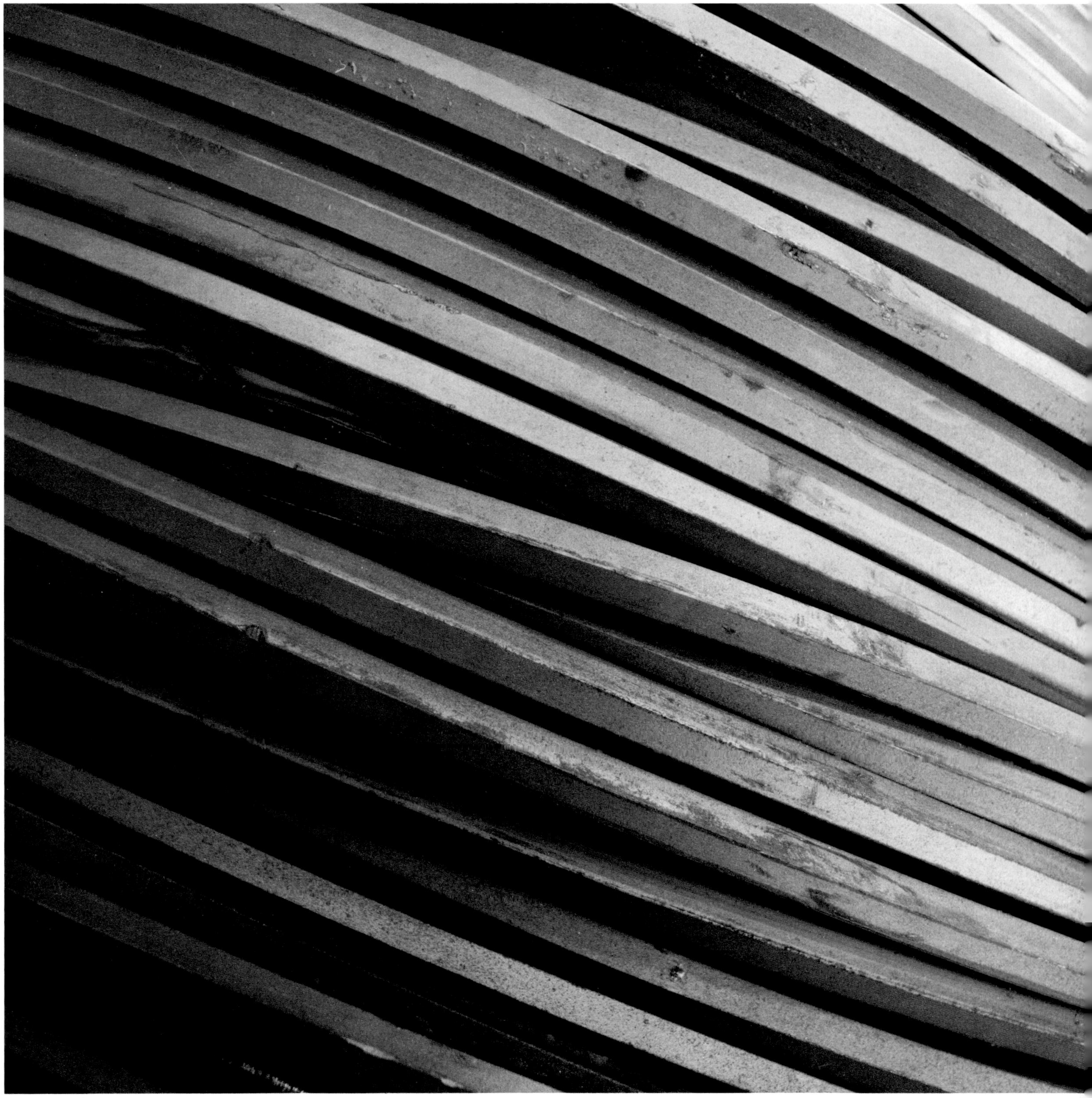

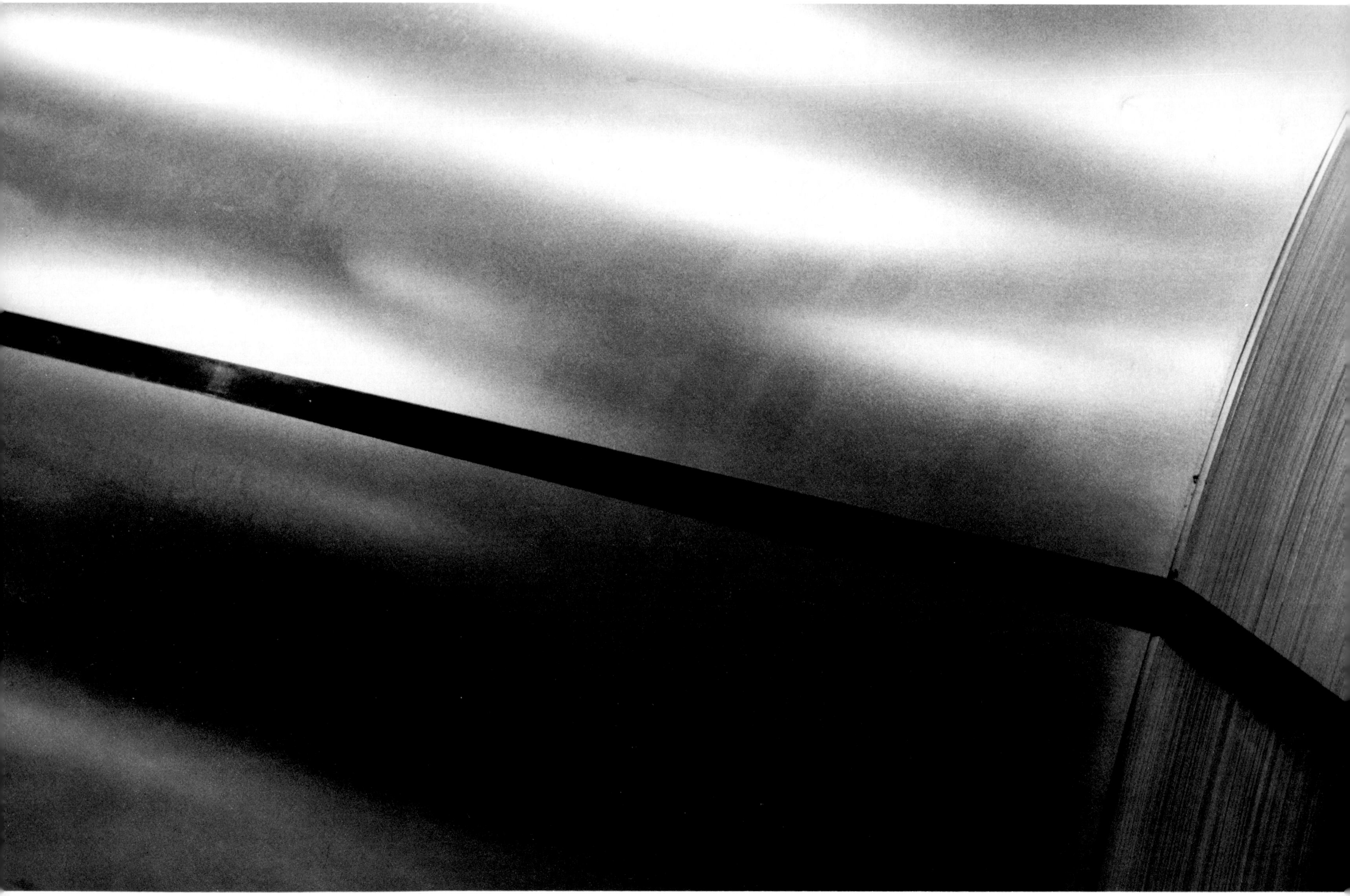

Coils of cold-rolled sheet steel.

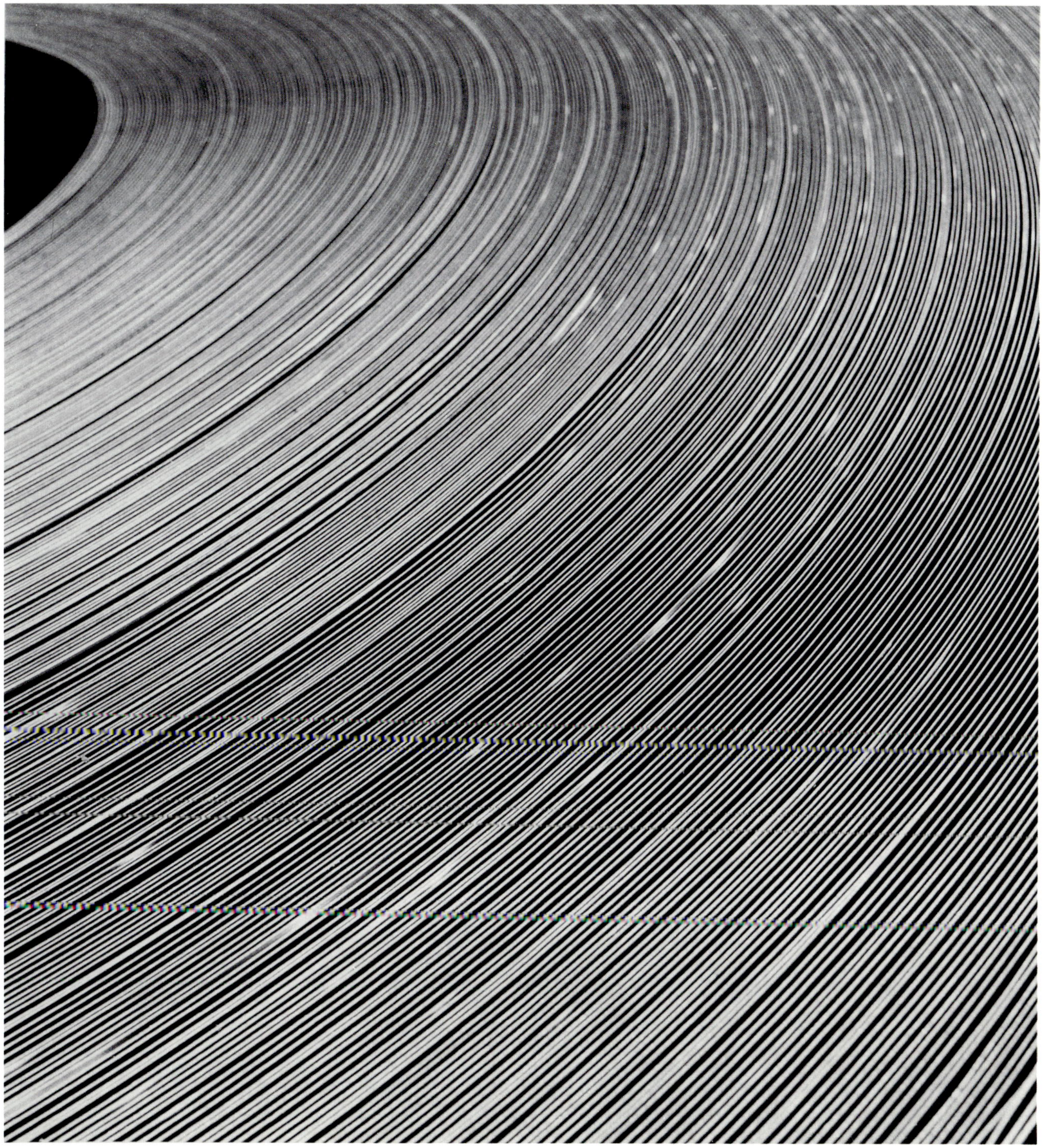